Communicating as Women in STEM

Communicating as Women in STEM

COMMUNICATING AS WOMEN IN STEM

CHARLOTTE BRAMMER
Samford University

Academic Press is an imprint of Elsevier
125 London Wall, London EC2Y 5AS, United Kingdom
525 B Street, Suite 1650, San Diego, CA 92101, United States
50 Hampshire Street, 5th Floor, Cambridge, MA 02139, United States
The Boulevard, Langford Lane, Kidlington, Oxford OX5 1GB, United Kingdom

Notices
Knowledge and best practice in this field are constantly changing. As new research and
experience broaden our understanding, changes in research methods, professional practices, or
medical treatment may become necessary.

Practitioners and researchers must always rely on their own experience and knowledge in
evaluating and using any information, methods, compounds, or experiments described herein.
In using such information or methods they should be mindful of their own safety and the
safety of others, including parties for whom they have a professional responsibility.

To the fullest extent of the law, neither the Publisher nor the authors, contributors, or editors,
assume any liability for any injury and/or damage to persons or property as a matter of
products liability, negligence or otherwise, or from any use or operation of any methods,
products, instructions, or ideas contained in the material herein.

Library of Congress Cataloging-in-Publication Data
A catalog record for this book is available from the Library of Congress

British Library Cataloguing-in-Publication Data
A catalogue record for this book is available from the British Library

ISBN: 978-0-12-802579-6

For information on all Academic Press publications visit our website at
https://www.elsevier.com/books-and-journals

Publisher: Mica Haley
Acquisition Editor: Mary Preap
Editorial Project Manager: Mary Preap
Production Project Manager: Poulouse Joseph
Designer: Victoria Pearson

Typeset by TNQ Technologies

CONTENTS

Introduction   *vii*

1. The Leaky Pipeline: Do We Have a (Communication) Problem?   **1**
Elizabeth's Story   14
Threshing It Out   21

**2. Stereotypes and Stacked Decks: Can Females *Really* Be
Scientists and Engineers?**   **23**
Anne's Story   28
Jessica's Story   28
Greta's Story   30
Jessica's Story   31
Karen's Story   32
Katie's Story   33
Tanya's Story   33
Juanita's Story   34
Jessica's Story   34
Threshing It Out   39

**3. My Team's Better Than Your Team: How Inclusive Is Office
Small Talk?**   **41**
Susan's Story   42
Amy's Story   43
Jeanna's Story, Information Sciences Graduate Student   45
Jessica's Story   45
Cara's Story   49
Amy's Story   49
Informants' Comments on Small Talk   50
Amy's Story   52
Threshing It Out   56

4. Power, Aggression, and Assertion: Who Gets to Speak?   **59**
Deborah's Story   59
Jade's Story   68
Ann's Story   68
Threshing It Out: Ann's Gems   69
Threshing It Out   70

5. **The Power of the Suit: Dressing for Success According to Whom?**   **71**
   Threshing It Out   77

6. **Communicating as Professionals in STEM: Some Closing Thoughts**   **79**
   Sadie's Story   80
   Mary's Story   82
   Jane's Story   82
   Lori's Story   83

References   *87*
Index   *95*

INTRODUCTION

To achieve project success in the construction field, my teams and I use several mantras. The most important one is "communicate early and often." This absolutely applies to everyday life personally and professionally. I have never run into a problem or issue that did not come down to some form of miscommunication or lack thereof. The point at which we think we are over communicating is when we are just beginning to communicate clearly with each other.

The textbook definition of communication is "a process by which information is exchanged between individuals through a common system of symbols, signs, or behaviors." The disconnect in communication between men and women in the workplace typically is rooted in not establishing that common system of behaviors to ensure we are communicating and leading to true understanding.

In recent years, women have taken a growing role in engineering, but gender stereotypes still exist in a male-dominated field. Ensuring that professional colleagues are not talking past one another and using a common system of expected behaviors is key to avoiding double standards based on gender and affording equal opportunities to achieve individual and collective potential. This concept should be foundational to any corporation's value proposition. Why would you want to hire anyone whom you did not expect to achieve their maximum potential and therefore your return on investment?

Two keys to increasing women's numbers in engineering are believed to be mentoring and dispelling myths and overly masculine perceptions of the field.

Girls often drop out of advanced math and science before the ninth grade. When they have strong female role models offering encouragement, the numbers rise. The media also portray women in traditional professions, and many young girls may not be aware of the vast opportunities in technical careers.

Women and girls continue to be significantly underrepresented in the STEM fields—a trend that starts early and comes at a serious cost to both the career prospects of our young women and the success of our economy. By ensuring that women and girls receive the exposure, encouragement, and support they need to enter, and succeed in, STEM fields, only then can we benefit from the full range and diversity of its talent.

Ten years ago, a member of the Alabama Public Service Commission asked Alabama Power what it was doing to reach out to young girls to spark an interest in STEM careers. The answer at the time was "not enough."

But that's hardly true today. Shortly after the question was raised, three other female employees and I got together and decided to look into the issue. After doing some research, we realized that the critical time to reach girls is in middle school—before students have to make a choice about challenging themselves in the advanced math and sciences in high school.

This is why we came up with the idea for iCan, which has grown into a hands-on career awareness and mentoring program that has since reached more than 1000 sixth- to eighth-grade girls in the Birmingham, Alabama area.

Our female engineering volunteers conduct multiple hands-on projects in the classroom during the school year, an annual field trip, and parent workshops to introduce them to the different types of engineering as well as the high-paying and fulfilling careers available in this field.

What we've learned is that in order to spark young girls' interest at the middle school level, they need to see and interact with role models. They need to see women who are doing these jobs, who have families, and who enjoy their work. They need to see that there are so many career options if they find math and science interesting. We show them how engineers help people, by designing things better to keep people safe during natural disasters, for example.

The program has been remarkably successful with the percentage of female students entering high school engineering academies increasing from as little as 6 to 21%. But the best measure of success is seeing the girls go on to pursue STEM studies after graduation. One of the first young ladies to go through the program went on to study electrical engineering in college, interned with Alabama Power, and has accepted a full-time position.

The keys to maintaining the pipeline of females once they enter into STEM fields is to ensure they receive equitable support and career advice regarding options in their field and coaching to achieve the advocacy necessary for career advancement. A common system of behaviors to encourage women to engage is key to identifying and closing blind spots and avoiding possible missed opportunities to innovate when women and men are communicating clearly and supporting each other.

I am proud to also live by another mantra, which is "don't complain about that which you can be part of the solution." It is critical that we achieve gender parity in the STEM fields to help solve those issues that are relevant to women and open up untapped market potential where everyone can benefit.

Jacquelyn O. Blakley
Director, EPC
Southern Power

CHAPTER 1

The Leaky Pipeline: Do We Have a (Communication) Problem?

Contents

Elizabeth's Story   14
Threshing It Out   21

The single biggest problem in communication is the illusion that it has taken place.

George Bernard Shaw

Because I am the only female in my department, I have to think about why I'm excluded. The men in my department go to lunch together every day. It's understood that I can barge in and go with them, but I'm not routinely asked. No one comes by my office or lab to say, 'hey, lunch time!'

Anonymous female chemist

We have a problem attracting and retaining women in STEM fields, especially in engineering and computer science, and the key question is simply *why*? STEM educators and others have worked on this frequently referenced "leaky pipeline" for years, and while they have made progress, they have not eliminated the gender imbalance in either academe or the workplace. Even in biology, perceived by some interested parties as having achieved the golden ring of gender equity, differences persist. Women are less likely to pursue post-doctoral study in biology, less likely to present at professional conferences, less likely to publish, and less likely to remain in academe despite having a majority presence and important exemplars in the discipline (e.g., Rachel Carson whose book *Silent Spring* brought

Communicating as Women in STEM
ISBN 978-0-12-802579-6
https://doi.org/10.1016/B978-0-12-802579-6.00001-X

1

widespread attention to ecology and Nobel winners Barbara McClintock and Linda B. Buck, to name only a few).

Yet, some question the existence of this leaky pipeline. Alexandra A. Killewald, Harvard sociologist, says, "A lot of things we think about in inequality of access to science are inequality to higher education" (Stix, 2012). A 2014 report from USA Today argues that the problem is one of excluding minorities from the hiring process rather than one of losing STEM students in the educational pipeline. Killewald and coauthor Yu Xie, in *Is American Science in Decline?*, argue that the leaky pipeline is inaccurate and unnecessarily pessimistic regarding the number of women and minorities in STEM fields. According to their data, more females switch into STEM majors than switch out.

Why does it matter? We seem to be in a cultural shift in which a lower percentage of males graduates from high school and fewer of those graduates go to college, while a higher percentage of females graduates from high school and chooses to attend college—but they are not heading into STEM. In raw numbers, the National Center for Education Statistics (NCES) reports that more women attend college than do men. More women attend as full-time students, and more women attend as part-time students. More than 57% of all undergraduate students in 2013 were female, and NCES projects this percentage to hold constant through 2024.

Women are not, however, flocking to STEM fields despite the higher incomes available to STEM graduates. Yet, our aging and increasingly technical society requires more individuals pursuing careers and research in STEM areas. Indeed, the President's Council of Advisors on Science and Technology calls for one million more STEM professionals in the United States, meaning the United States must "increase the number of students who receive undergraduate STEM degrees by about 34% annually over current rates" (2012). Thus, at a time when we need more students going into STEM fields, the traditional pool of students who choose these fields is not growing (Table 1.1).

The National Science Foundation reports that by 2010, women reached numeric parity with men in social sciences as well as biosciences; however, women are underrepresented in physical sciences, mathematics, computer science, and engineering. This disparity is

Table 1.1 Bachelor's, master's, and doctor's degrees conferred by postsecondary institutions, by sex of student and discipline division: 2014—15

Discipline Division	Bachelor's Degrees			Master's Degrees			Doctor's Degrees[a]		
	Total	Males	Females	Total	Males	Females	Total	Males	Females
All fields, total	**1,894,934**	**812,669**	**1,082,265**	**758,708**	**306,590**	**452,118**	**178,547**	**84,921**	**93,626**
Agriculture and natural resources	36,277	17,585	18,692	6426	2904	3522	1561	811	750
Architecture and related services	9090	5116	3974	8006	4147	3859	272	145	127
Area, ethnic, cultural, gender, and group studies	7782	2293	5489	1847	666	1181	312	120	192
Biological and biomedical sciences	109,896	45,102	64,794	14,650	6249	8401	8053	3763	4290
Business, management, marketing, and personal and culinary services	363,799	191,310	172,489	185,222	98,587	86,635	3116	1716	1400
Communication and communications technologies	95,785	35,218	60,567	10,135	3111	7024	644	259	385
Computer and information sciences and support services	59,581	48,840	10,741	31,474	21,892	9582	1998	1548	450
Education	91,623	18,473	73,150	146,561	33,980	112,581	11,772	3838	7934

Continued

Table 1.1 Bachelor's, master's, and doctor's degrees conferred by postsecondary institutions, by sex of student and discipline division: 2014—15—cont'd

Discipline Division	Bachelor's Degrees			Master's Degrees			Doctor's Degrees[a]		
	Total	Males	Females	Total	Males	Females	Total	Males	Females
Engineering and engineering technologies	115,096	93,532	21,564	51,439	38,452	12,987	10,362	7958	2404
English language and literature/letters	45,847	14,053	31,794	8928	3052	5876	1418	554	864
Family and consumer sciences/human sciences	24,584	3012	21,572	3148	429	2719	335	79	256
Foreign languages, literatures, and linguistics	19,493	6067	13,426	3566	1241	2325	1243	532	711
Health professions and related programs	216,228	33,658	182,570	102,897	18,780	84,117	71,003	29,426	41,577
Homeland security, law enforcement, firefighting, and related prot. services	62,723	33,640	29,083	9643	4782	4861	193	96	97
Legal professions and studies	4420	1348	3072	7924	3688	4236	40,329	20,964	19,365
Liberal arts and sciences, general studies, and humanities	43,647	16,136	27,511	2794	999	1795	96	48	48
Library science	99	16	83	5259	1025	4234	44	11	33
Mathematics and statistics	21,853	12,462	9391	7589	4508	3081	1801	1298	503
Military technologies and applied sciences	276	225	51	71	44	27	0	0	0

Field of study									
Multi/interdisciplinary studies	47,556	15,930	31,626	8098	3145	4953	840	380	460
Parks, recreation, leisure, and fitness studies	49,006	25,948	23,058	7639	4329	3310	311	184	127
Philosophy and religious studies	11,072	7004	4068	1912	1222	690	762	522	240
Physical sciences and science technologies	30,038	18,478	11,560	7100	4438	2662	5823	3826	1997
Precision production	48	33	15	4	3	1	0	0	0
Psychology	117,557	26,802	90,755	26,773	5538	21,235	6583	1624	4959
Public administration and social service professions	34,363	6145	28,218	46,083	11,187	34,896	1123	374	749
Social sciences and history	166,944	85,478	81,466	20,533	10,318	10,215	4828	2589	2239
Theology and religious vocations	9708	6612	3096	14,271	9437	4834	1927	1415	512
Transportation and materials moving	4711	4133	578	960	794	166	5	4	1
Visual and performing arts	95,832	38,020	57,812	17,756	7643	10,113	1793	837	956
Not classified by field of study	0	0	0	0	0	0	0	0	0

NOTE: Data are for postsecondary institutions participating in Title IV federal financial aid programs. Aggregations by field of study derived from the Classification of Instructional Programs developed by the National Center for Education Statistics.

[a]Includes PhD, EdD, and comparable degrees at the doctoral level. Includes most degrees formerly classified as first professional, such as MD, DDS, and law degrees.

Source: U.S. Department of Education, National Center for Education Statistics, Integrated Postsecondary Education Data System (IPEDS), Fall 2009 through Fall 2015. (2016). Retrieved from https://nces.ed.gov/programs/digest/d16/tables/dt16_318.45.asp.

particularly pronounced in computer science and engineering where women comprise less than 30% of undergraduate students. The disparity improves at the graduate level for STEM fields, with 42% of doctorates in these fields claimed by women; the percentage, however, has held fairly constant since then despite the fact that 57% of all undergraduates are women. Moreover, since 2004, more female US citizens and permanent residents have earned doctorates (across all fields) than have males in the same category. The curves appear to have flattened since 2012 (Tables 1.2 and 1.3).

STEM fields, perhaps especially computer science and engineering, are gendered male, not necessarily by some misogynistic design but because historically they have been largely populated by males. We tend to think of organizations as gender neutral, but as Acker posited in 1990, they are not. Thinking about organizations as gendered entities moves us from accepting traditional (gendered male) patterns of interaction and structure as normal and toward recognizing the inherent and persistent frames of "advantage and disadvantage, exploitation and control, action and emotion, meaning and identity" they contain. Acker (1990) acknowledges five processes of gendering within organizations. Let's discuss them briefly here because they resonate with the women's stories that I heard and share in this book.

First, Acker (1990) identifies how organizations structure jobs, location, and power along gendered lines and roles. For example, men's work generally means skilled work; women's, unskilled. Manager means male and privileges traditional male behaviors. Engineers are stereotypically males; administrative assistants are females. Second, language, symbols, and images of those structural divisions reinforce those lines. To get an idea of this concept, take a moment to imagine or describe what comes to mind when you think of an English teacher. What does this person look like, sound like, etc. Really get an image of this person in your mind. Got it? Now, think about an English professor. Describe how this person looks and sounds. How does the professor compare with the teacher? Did you imagine a middle-aged (or older) woman, probably unmarried, as the teacher? Did the professor wear a tweed jacket with leather-patch

Table 1.2 Number of STEM degrees conferred by US postsecondary institutions by sex of student: 2008–09 through 2014–15

	Year	Bachelor's Degrees	Master's Degrees	Doctor's Degrees	All Degrees
Total	2008–09	243,031	77,006	23,654	343,691
	2009–10	253,650	79,781	23,705	357,136
	2010–11	267,480	86,271	24,587	378,338
	2011–12	286,788	91,612	25,528	403,928
	2012–13	302,340	95,375	26,577	424,292
	2013–14	318,612	100,078	28,070	446,760
	2014–15	335,837	112,252	28,037	476,126
Males	2008–09	157,319	53,411	15,511	226,241
	2009–10	164,612	55,217	15,363	235,192
	2010–11	173,493	59,617	16,203	249,313
	2011–12	185,802	63,212	16,685	265,699
	2012–13	196,343	65,369	17,412	279,124
	2013–14	206,935	67,904	18,468	293,307
	2014–15	217,830	75,539	18,393	311,762
Females	2008–09	85,712	23,595	8143	117,450
	2009–10	89,038	24,564	8342	121,944
	2010–11	93,987	26,654	8384	129,025
	2011–12	100,986	28,400	8843	138,229
	2012–13	105,997	30,006	9165	145,168
	2013–14	111,677	32,174	9602	153,453
	2014–15	118,007	36,713	9644	164,364

Source: U.S. Department of Education, National Center for Education Statistics, Integrated Postsecondary Education Data System (IPEDS), Fall 2009 through Fall 2015. Modified by Brammer, March 2018.

Table 1.3 National science foundation enrollment status of S&E graduate students, by field and sex: 2014

Field	All S&E Graduate Students			Female			Male		
	Total	Full Time	Part Time	Total	Full Time	Part Time	Total	Full Time	Part Time
S&E	601,883	447,096	154,787	253,493	185,940	67,553	348,390	261,156	87,234
Science	437,395	322,714	114,681	213,803	154,746	59,057	223,592	167,968	55,624
Agricultural sciences	17,505	12,319	5186	8968	6431	2537	8537	5888	2649
Biological sciences	78,490	64,638	13,852	44,593	35,894	8699	33,897	28,744	5153
Anatomy	554	519	35	280	262	18	274	257	17
Biochemistry	5025	4592	433	2423	2206	217	2602	2386	216
Biology	16,697	12,255	4442	9340	6664	2676	7357	5591	1766
Biometry and epidemiology	8326	6474	1852	4915	3764	1151	3411	2710	701
Biophysics	877	849	28	306	298	8	571	551	20
Botany and plant biology	1877	1669	208	938	827	111	939	842	97
Cell and molecular biology	6214	5819	395	3389	3152	237	2825	2667	158
Ecology	1416	1032	384	756	563	193	660	469	191
Entomology and parasitology	1304	1106	198	651	557	94	653	549	104
Genetics	2411	2343	68	1425	1385	40	986	958	28
Microbiology, immunology, and virology	4670	4345	325	2658	2469	189	2012	1876	136

Nutrition	5868	3780	2088	4730	3040	1690	1138	740	398
Pathology	1036	1013	23	612	598	14	424	415	9
Pharmacology and toxicology	2916	2594	322	1619	1427	192	1297	1167	130
Physiology	3415	3158	257	1766	1608	158	1649	1550	99
Zoology	1093	708	385	693	395	298	400	313	87
Biosciences nec	14,791	12,382	2409	8092	6679	1413	6699	5703	996
Communication	11,942	7544	4398	7804	4859	2945	4138	2685	1453
Computer sciences	76,546	52,069	24,477	21,038	14,509	6529	55,508	37,560	17,948
Earth, atmospheric, and ocean sciences	15,710	12,462	3248	7144	5726	1418	8566	6736	1830
Atmospheric sciences	1466	1262	204	538	481	57	928	781	147
Geosciences	8821	6911	1910	3630	2893	737	5191	4018	1173
Ocean sciences	2666	2169	497	1528	1227	301	1138	942	196
Earth, atmospheric, and ocean sciences nec	2757	2120	637	1448	1125	323	1309	995	314
Family and consumer sciences and human sciences	4302	2577	1725	3568	2122	1446	734	455	279
Mathematics and statistics	25,874	20,445	5429	9234	7020	2214	16,640	13,425	3215

Continued

Table 1.3 National science foundation enrollment status of S&E graduate students, by field and sex: 2014—cont'd

Field	All S&E Graduate Students			Female			Male		
	Total	Full Time	Part Time	Total	Full Time	Part Time	Total	Full Time	Part Time
Mathematics and applied mathematics	18,863	14,785	4078	6176	4482	1694	12,687	10,303	2384
Statistics	7011	5660	1351	3058	2538	520	3953	3122	831
Multidisciplinary and interdisciplinary studies	7196	4862	2334	3445	2322	1123	3751	2540	1211
Neurobiology and neuroscience	4923	4703	220	2566	2416	150	2357	2287	70
Physical sciences	40,332	36,266	4066	13,118	11,638	1480	27,214	24,628	2586
Astronomy and astrophysics	1241	1219	22	427	417	10	814	802	12
Chemistry	22,936	20,565	2371	9352	8268	1084	13,584	12,297	1287
Physics	15,564	14,047	1517	3128	2801	327	12,436	11,246	1190
Physical sciences nec	591	435	156	211	152	59	380	283	97
Psychology	48,833	34,683	14,150	36,243	25,364	10,879	12,590	9319	3271
Clinical psychology	8781	7057	1724	6783	5435	1348	1998	1622	376
Psychology, general	12,643	9902	2741	8710	6760	1950	3933	3142	791
Psychology nec	27,409	17,724	9685	20,750	13,169	7581	6659	4555	2104
Social sciences	105,742	70,146	35,596	56,082	36,445	19,637	49,660	33,701	15,959
Agricultural economics	1931	1598	333	822	704	118	1109	894	215

Anthropology	7955	6248	1707	5126	4041	1085	2829	2207	622
Economics (except agricultural)	14,604	12,426	2178	5267	4508	759	9337	7918	1419
Geography	4810	3354	1456	2133	1520	613	2677	1834	843
History and philosophy of science	371	332	39	189	166	23	182	166	16
Linguistics	3489	2668	821	2136	1574	562	1353	1094	259
Political science and government	47,370	27,478	19,892	24,877	14,138	10,739	22,493	13,340	9153
Sociology	8637	6590	2047	5402	4009	1393	3235	2581	654
Sociology and anthropology	197	120	77	132	83	49	65	37	28
Social sciences nec	16,378	9332	7046	9998	5702	4296	6380	3630	2750
Engineering	164,488	124,382	40,106	39,690	31,194	8496	124,798	93,188	31,610
Aerospace engineering	5116	3907	1209	775	596	179	4341	3311	1030
Agricultural engineering	1740	1412	328	704	580	124	1036	832	204
Architecture	1817	1588	229	888	766	122	929	822	107
Biomedical engineering	9510	8336	1174	3810	3354	456	5700	4982	718
Chemical engineering	9870	8629	1241	3107	2728	379	6763	5901	862
Civil engineering	20,789	15,219	5570	6093	4635	1458	14,696	10,584	4112
Electrical engineering	51,909	39,868	12,041	10,961	8763	2198	40,948	31,105	9843

Continued

Table 1.3 National science foundation enrollment status of S&E graduate students, by field and sex: 2014—cont'd

Field	All S&E Graduate Students			Female			Male		
	Total	Full Time	Part Time	Total	Full Time	Part Time	Total	Full Time	Part Time
Engineering science, mechanics, and physics	2162	1575	587	429	324	105	1733	1251	482
Industrial and manufacturing engineering	14,845	9106	5739	3824	2417	1407	11,021	6689	4332
Mechanical engineering	25,651	19,015	6636	3772	2894	878	21,879	16,121	5758
Metallurgical and materials engineering	7518	6625	893	2181	1941	240	5337	4684	653
Mining engineering	396	307	89	76	54	22	320	253	67
Nuclear engineering	1467	1201	266	221	181	40	1246	1020	226
Petroleum engineering	2056	1662	394	414	350	64	1642	1312	330
Engineering nec	9642	5932	3710	2435	1611	824	7207	4321	2886

nec, not elsewhere classified; *S&E*, science and engineering.

NOTES: In 2014, the survey frame was updated following a comprehensive frame evaluation study. The study identified potentially eligible but not previously surveyed US academic institutions with master's- or doctorate-granting programs in science, engineering, or health. A total of 151 newly eligible institutions were added, and two private for-profit institutions offering mostly practitioner-based graduate degrees were determined to be ineligible.

Source: National Science Foundation, National Center for Science and Engineering Statistics, Survey of Graduate Students and Postdoctorates in Science and Engineering, 2014.

elbows? Was the professor you imagined male? Was marital status part of the picture? How do you envision an anesthesiologist? A nurse anesthetist? Did the term *nurse* gender the role female where the first was gendered male? Why?

A third aspect of gendering within organizational structure relates to patterns of interaction between/among organizational members, and this aspect is particularly salient for this book. Consider for a moment your experiences in interacting with other members of your gender and then consider the differences when you interact with another gender. Interactions between men, between women, and between women and men differ in terms of interruptions, turn-taking, topic setting, and conversational flow, among other ways. Frequently, men are perceived as the actors, the ones doing things and controlling the interaction, while women are expected to provide emotional support. These behaviors are part of self-presentation, a fourth aspect of organizational gendering, and males and females are expected to act in ways that reflect societal and organizational norms. Adhering to such societal norms is deemed appropriate and polite by others; failing to adhere to the norms is penalized. Women tend to be penalized more frequently and more harshly.

Finally, gender is embedded in organizational logic, as seen in written work rules, labor contracts, managerial directives, job evaluations, etc. The so-called unencumbered white male is privileged over others, especially women, because the rules are written to favor those "committed to paid employment [as] more suited to responsibility and authority" while women, who "must divide their commitments" between public/organizational and domestic work, are disadvantaged. The real problem, as Acker (1990, p. 152) notes:

> ... the abstract worker is actually a man, and it is the man's body, its sexuality, minimal responsibility in procreation, and conventional control of emotions that pervades work and organizational processes. Women's bodies—female sexuality, their ability to procreate and their pregnancy and breast-feeding, and childcare, menstruation, and mythic emotionality—are suspect, stigmatized.

Each of these aspects of gendering burgeons in organizations and disciplines that have reflected these norms.

Culturally, STEM fields are not supposed to be attractive to females. Gendering—learning how to be girls and boys—starts early, often before birth. As soon as a mother learns the biological sex of her baby, she begins thinking about names, planning the nursery, and buying clothing to match the child's biologically identified gender. Infant girls begin receiving dolls before they are even born; boys get dump trucks and action figures. In school, girls are expected to perform better on verbal tasks and to be less active; boys are expected to perform better in science and math and to be more active (and are much more likely to be identified with having attention deficit and hyperactivity issues). Girls are supposed to be polite, obedient; boys are expected to be rowdier, louder, and even less obedient at times. Children who don't conform to these gender stereotypes are often ostracized and teased, even bullied by their peers. Building things, dissecting things, gaming, and programming are considered masculine activities. Girls are supposed to be impressed by what boys build and repulsed by dissected carcasses. These gendered behaviors as well as other cultural expectations or norms are perpetuated through communication, verbally and nonverbally.

Several female engineers, when discussing their undergraduate experience, noted how frequently professors used car analogies in effort to clarify concepts. While these analogies seemed to work for male students, they did not clarify anything for the females. Most female engineers, particularly those in mechanical, civil, and electrical disciplines, reported being either the only female or one of three females in their engineering courses.

ELIZABETH'S STORY

As an undergraduate, I was usually either the only female, or if I was lucky, I would be one of 3 out 30–40 in a class. In lots of classes, the male faculty—and they were all males—would use car analogies. I didn't get the analogies, so I would leave class and try to find other analogies to help me understand the concepts. My professors related to most of the class, but not to me.

The norm of using automobiles mechanics or maintenance makes sense if professors are working under the assumption that everyone in the class is interested in and knowledgeable about how they work, and this assumption feeds from the stereotype that engineers—traditionally gendered male—obviously work on cars and other types of machinery. This assumption may be wrong for both the females and males in the classroom, excluding females who generally comply with their traditional gendered interests and males who do not comply or do not show interest in car maintenance and mechanics.

A 2013 study of commonly used K–12 science textbooks found biases against women and minorities, and these biases were greatest in the lower K–3 texts. As Blumberg writes in 2015 UNESCO report, "Their domination of instruction, coupled with their pervasive gender stereotyping and underrepresentation of females, result in textbooks often being cited as limiting girls' academic achievements and adult options" (p. 1). Blumberg recommends revising textbooks for inclusion and training teachers to mitigate the negative impacts.

The relationship between culture and communication is intimate, even somewhat symbiotic. Culture influences what is perceived and valued, and these perceptions and values are shared through communication via storytelling, language, and social norms that guide appropriateness of speech, behaviors, and interaction. Storytelling serves many purposes, including teaching others what to pay attention to and what matters. For example, the Ida V. Moffett School of Nursing at Samford University uses Mrs. Moffett's story to convey that nursing is more than a job or profession: it is a "calling," a "mission." Mrs. Moffett is praised for having a "devout character and never-ending sense of compassion [through which she] touched the hearts and lives of countless patients and professionals" (https://www.samford.edu/nursing/mrs-moffetts-legacy). Notably, she is always referenced as *Mrs.* Moffett to show respect to this hero. A nurse who cared for Mrs. Moffett during the last stage of her life told of buying a traditional white nursing hat to show respect for this "amazing woman who taught us what it means to be a nurse." Service is central to the school's nursing mission.

Compare that story to this brief statement on Cal Tech's website: "Caltech's founding fathers—astronomer George Ellery Hale, physicist Robert Andrews Millikan, and chemist Arthur Amos Noyes—were nicknamed 'Tinker, Thinker, and Stinker.'" The stories about engineers tend to focus on discoveries and accomplishments sans human impact, yet those same discoveries and accomplishments have made tremendous impacts on how people live. Think about how much our lives have been impacted by Thomas Edison and Steve Jobs or Bill Gates. The difference in focus reflects traditional gender roles and ideals.

We gain deeper knowledge of culture through looking at stories as well as by studying language. Thus, most high school and college students study foreign languages, and businesses value those fluent in multiple languages. Yet, we also learn about our own cultures by studying our language and how we use it. Linguists George Lakoff and Mark Johnson's *Metaphors We Live By*, originally published in 1980 and now recognized as a classic, explains how metaphors are much more than literary devices or rhetorical tropes. In their words, metaphors are cognitive constructs that both influence and reflect how we think, and they explain their understanding of metaphors through an extended explication of how westerners, particularly in the United States, conceive of *argument* as *war* and convey it, perhaps unwittingly, through words, such as

> *He attacked every weak point in my argument.*
> *His criticisms were right on target.*
> *I demolished his argument.*
> *I've never won an argument with him.*
> *You disagree? Okay, shoot!*
> *If you use that strategy, he'll wipe you out.*
> *He shot down all of my arguments. (Lakeoff and Johnson, p. 4)*

Through such language, the concept of argument as war is shared and promulgated. We are still learning how much language matters in shaping our thoughts, as Thibodeau and Boroditsky (2011) demonstrate. In a series of five experiments, they manipulated the use of two common metaphors used to describe crime (e.g., crime is a beast/virus) and found "that people chose information that was likely to confirm and elaborate the bias suggested by the metaphor — an effect that persisted even when people were presented with a full set of possible solutions." In plain language, our thoughts are influenced by language so much that we seek evidence to confirm the bias it contains.

Through language use as well as nonverbal communication, we learn social norms. We learn how to exchange greetings and who is expected to speak first. We learn how to shake hands and the etiquette that goes with it. In some parts of the United States for example, women are still expected to initiate handshakes if they want to participate in this ritualistic greeting. Similarly, in the southern part of the United States, men are expected to hold doors open for women. These nonverbal behaviors bleed into expectations in the workplace, and sometimes they create ambiguities and awkward interactions. For example, if a man has been brought up to hold doors for women and he continues this practice in the workplace, a female peer may perceive it as demeaning or patriarchal rather than as politeness, as a sign that she is not considered an equal. Similarly, if he fails to hold the door, he may be perceived as rude, lacking manners and basic courtesy. Such behaviors undergo scrutiny and sometimes create uncomfortable and awkward interactions, and both males and females are challenged by changing social practice. Importantly, because males have traditionally held more power relative to women's power, subtleties in social norms can exclude women, discriminate against them, or simply make them question whether they belong in that particular place—e.g., an engineering classroom.

Gender roles, both their negative and positive aspects, are deeply embedded in society and are central to self-identification and perception. Gender theorists, psychologists, and sociologists offer descriptions and criticisms of gender-based stereotyping, extending their discussions from women to the LGBTQ community and finally to men. Many individuals, women, men, and LGBTQ individuals have collectively fought for equity in social and professional realms, and they have made significant progress in some areas. Yet, inequities persist. Culture changes slowly. Thus, the female chemist, quoted at the beginning of this chapter, interrogates her perceptions as well as her peers' behaviors to understand why the men comfortably stop by each other's work spaces to notify each other of their routine lunch date, yet rarely stop by hers. If she attends *their* informal lunch breaks, she must go of her own volition, without the expressed invitation that other (all male) department members receive. While she believes this exclusion is unintentional, citing the overall good rapport among her department, I wonder to what extent the exclusion results from enacted gender roles. In 2000, linguist Su Olsson documents through women's workplace narratives that their biggest challenges come from how others, especially but not exclusively men, treat them. Gendered behaviors and language often operate beneath conscious thought, as automatic responses and actions.

We use language to construct gender and social norms, and linguists frequently parse such language use in terms of discursive engagement, discursive content, and personas (McGonnell-Ginet offers extensive discussion of these concepts in much of her work. See for example, *Gender, Sexuality, and Meaning: Linguistic Practice and Politics, 2011*). Let's briefly discuss each of those concepts as a way of enlightening the approach used in this book. Discursive engagement involves the pragmatics of interaction such as how turn-taking happens within a conversation, who is expected to speak first and for how long, and how interruptions (and by whom) are made and accepted—or not. Numerous studies have demonstrated how men frequently interrupt or overtalk women in the workplace with impunity; women who interrupt, however, are not generally well regarded, are viewed as pushy—or worse. We'll delve more deeply into this conundrum later.

Speech, discursive content, sometimes conveys unchallenged assumptions, stereotypes, and biases. For example, we frequently specify gender when it isn't necessary: a lady doctor, a female engineer, a male nurse, or male kindergarten teacher, etc. Such delineation implies something unusual about the particular person in that role: women aren't surgeons or engineers and men aren't nurses and kindergarten teachers. Thus, individuals who work in fields outside the perceived gender norm must somehow come to terms with being perceived as anomalies, and they do so in a variety of ways, sometimes apologetically and sometimes defiantly. How they come to terms with those roles, specifically how they portray their gender in such roles, signals their personas. Gender is always present in how we act and interact.

As I was drafting this book, James Damore published his lengthy screed against Google's and society's efforts to recruit and retain women in STEM. His "Google's Ideological Echo Chamber" is perhaps more convincing, albeit unintentionally so, than I have been in positing why we need to address the leaky pipeline. His belief that women somehow are not interested in STEM because they are women belies the deep cultural constraints that discourage them. Damore's message and its popularity among a significant number of people, not to mention the publicity he received, conveys how widespread this belief is, how ingrained it is, and how emotionally and socially challenging such fields are for women. YouTube CEO and Google employee Susan Wojcicki references her own experience in technology in Forbes:

> *Yesterday, after reading the news, my daughter asked me a question. "Mom, is it true that there are biological reasons why there are fewer women in tech and leadership?"*
>
> *That question, whether it's been asked outright, whispered quietly, or simply lingered in the back of someone's mind, has weighed heavily on me throughout my career in technology. Though I've been lucky to work at a company where I've received a lot of support—from leaders like Larry Page, Sergey Brin, Eric Schmidt, and Jonathan Rosenberg to mentors like Bill Campbell—my experience in the tech industry has shown me just how pervasive that question is.*

Damore is not the first and probably will not be the last male to suggest that women are incapable of some job or activity or intellectual pursuit by virtue of being a woman. It's actually a fairly common and well-worn trope. I'm always curious, however, as to what motivates someone to reframe it every few years. I've heard a male faculty member tell a class of graduate students—five females, one male—that "*girls* get good grades, but it's the men who earn the honors and come up with the great ideas." Some may choose to deny the reality that women are interested in STEM, and women and girls are, and more can be, successful in STEM. Biological sex does not determine academic interest and curiosity.

This book is about how communication practices, particularly gendered practices, may impede the inclusion and progress of women (and others) in STEM fields, how practitioners and mentors can address communication and disciplinary acculturation, and how changes in communication practice can lead to cultural changes that are more inclusive. Becoming mindful of cultural differences and adapting or adding communicative styles are helpful strategies for successfully navigating such differences. By scaffolding learning about cultural and gender differences in communication, we enable women, minorities, and concerned faculty and administrators to develop broader understanding of how communication happens and how to question their perceptions with hopes of promoting perseverance in the disciplines. By making explicit some gendered communication practices, we also hope to assist individuals, especially those in positions to mentor and encourage women and minorities in STEM disciplines, to identify practices that exclude, however, unintentionally, in order to broaden their communication styles toward inclusion. Yet, we must also be mindful that some individuals intentionally harass women and minorities, and such individuals must be held accountable. Women and minorities who are targeted need support from peers and supervisors to deal with workplace bullies.

In gathering information for this book, I was fortunate to interview 49 women in various STEM disciplines. While a few work in academe, most participants are gainfully employed in STEM professions as engineers, information technologists, chemists, technical sales, etc. Some are actively involved in community service to engage young girls and/or minorities in STEM fields. They are giving back even as some described being discouraged by older women in their fields. They are observant and committed to their disciplines. They ranged in age from their early twenties to their late fifties. These women represent a geographically diverse sampling. After talking with one woman, I was frequently referred to a friend or former classmate in another state. I met women in coffee shops and restaurants and spoke with some over the phone. To protect their identities, I refer to them by profession or pseudonyms. Their stories are unique yet share common threads of success, exclusion, exception, and second-guessing. From their stories, we see examples of how communication practices in STEM disciplines may discourage or even disenfranchise women and others. Importantly, we also see opportunities to make these fields more open.

THRESHING IT OUT

How have you heard the gender disparity in STEM fields explained or excused?

What is your response to Damone? To what extent do you accept or agree with his points? How might you respond to a peer, a faculty member, or supervisor who espouses views similar to Damone's? If you are a mentor, what sorts of views, prejudices, and myths persist in your field that discourage and/or disparage females?

Brainstorm some ways to address these exclusionary views and myths in your local group. Gather information by asking the women around you about their experiences with these issues. Talk with young women in your fields and convey support by acknowledging the unfounded views and encouraging the women to persist as you work with them to change the culture.

CHAPTER 2

Stereotypes and Stacked Decks: Can Females *Really* Be Scientists and Engineers?

Contents

Anne's Story	28
Jessica's Story	28
Greta's Story	30
Jessica's Story	31
Karen's Story	32
Katie's Story	33
Tanya's Story	33
Juanita's Story	34
Jessica's Story	34
Threshing It Out	39

> *It takes a near act of rebellion for even a four-year-old to break away from society's expectations.*
> **Sheryl Sandberg, Lean In: Women, Work, and the Will to Lead**

Anyone familiar with the popular sitcom, *The Big Bang Theory*, can probably identify the STEM stereotypes: super smart, socially awkward, male, white, Asian, or Indian. Sheldon Cooper conveys the full stereotype. The show also features the brilliant, socially awkward female, Amy Farrah Fowler as Sheldon's counterpart. She is a rarity in the masculine world of STEM. Penny, the quintessential "dumb blonde" on the show is sexy and culturally savvy even as she struggles with paying bills instead of buying more stylish shoes. In season four of the show, Amy Farrah Fowler develops somewhat of a crush on Penny because she wants to learn to be as sexy as Penny is. Despite her advanced degrees and academic accomplishments, Amy longs for the stereotypic female role.

Communicating as Women in STEM
ISBN 978-0-12-802579-6
https://doi.org/10.1016/B978-0-12-802579-6.00002-1

23

We could get bogged down into a chicken or egg conundrum here, but how these stereotypes developed is less immediate than how they impact our interactions and how we change them.

What are stereotypes? We have general ideas that stereotypes are terrible things, and in many respects, they are terrible. Yet, stereotypes also help us categorize and process massive amounts of information in a relatively short period of time so that we avoid constant cognitive overload. Formally, stereotypes are a collection of generally held beliefs about a group or "type" of people, animals, situations, and things. For example, many people hold very negative views of pit bulls as a dog breed and stereotype all pit bulls as aggressive and unpredictable. Media feeds this stereotype by conveying stories of pit bulls whose actions support the stereotype, and even if the media conveys a story of a pit bull not conforming to the scary role, the incident is framed as showing aberrant rather than normal behavior for the breed. Labrador retrievers, however, are stereotyped as the perfect family dog— friendly, goofy, playful, and protective. The general beliefs about the two breeds have some factual support: between 2005 and 2016, pit bulls (and pit bull mixes) are blamed for 254 (65%) fatal dog attacks (other statistics available at http://dogsbite.org). We will leave the discussion of what happened during each of these attacks for others to sort out, but we probably know some pit bulls that are very friendly and some labs that are overly aggressive. Veterinarians and training experts often cite training and nurturance as important influences on animal behavior. My point here is not to delve into the dog breed controversy but rather to point out that stereotypes must be interrogated rather than being allowed to control our thoughts. To suggest that a group of people, based on gender, ethnicity, or any other categorization, can or cannot do a certain job or engage a certain discipline or will or will not be interested in pursuing certain professions reflects an uncritical acquiescence to stereotype.

Stereotypes, left unexamined, lead to bias and discrimination, which is the terrible side of stereotypes. If we assume that all labs are predisposed to being friendly and goofy, we will likely interact with our neighbor's lab as if we know it, petting it, getting closer to it, etc. Our reaction may be quite different if we discover our neighbor has a pit bull, even if we have no indication that particular dog is aggressive and instead see the dog playing with the neighbor's children and rolling on its back for submissive tummy rubs. Perhaps our prejudiced thoughts will keep us from being bitten, but they will surely also inhibit our ability to develop a relationship with either our neighbor or the pet. Acting on our biased thoughts also impacts the objects of those thoughts, creating or denying opportunities.

We continue to experience long-term impacts of denied opportunities for some people groups. The American Civil Rights Movement of the 1950s and 1960s sought remedy for institutionalized injustices, including separate and unequal public schools, lack of access to higher education, exclusion from professional organizations and employment opportunities, to list only a few. Yet, many urban schools that serve students predominantly from minority communities remain underfunded. Disparities in performance on standardized tests, including the newly revised American College Testing (ACT), persist. The 2017 ACT national profile (Table 2.1) report reveals persistent gaps between the performance of white Americans and minorities, with the notable exception of test takers who self-identify as Asian. African Americans' average composite score in 2013 was 16.9; by 2017, it rose to just 17.1. Whites scored 22.2 and 22.4 during those same years. For the test dates reported, more females (52%) than males (46%) took the ACT, and the composite scores for both genders were nearly equal: 21.1 for females, 21.0 for males. Yet, males scored 21.5 compared to 20.8 for females in the STEM categories, with females apparently losing the most ground in science. The Civil Rights Movement and subsequent legislation provide legal recourse, but they have not eliminated social injustice and the long-term effects of bias and discrimination.

Not all bias is external. As we accept societal expectations and limitations of our own abilities and potential, we become our own stumbling blocks. Stereotype threat, identified by Steele and Aronson (1995), is the "risk of confirming, as self-characteristic, a negative stereotype of one's group." In other words, we all look at the world through glasses that are designed, shaped, and colored by others. These frames and lenses reflect the stereotypes we have internalized,

Table 2.1 Five year trends—percent and average composite score by race/ethnicity

Race/Ethnicity	2013			2014			2015			2016			2017		
	N	%	Avg	N	%	Avg	N	%	Avg	N	%	Avg	N	%	Avg
All Students	1,799,243	100	20.9	1,845,787	100	21.0	1,924,436	100	21.0	2,090,342	100	20.8	2,030,038	100	21.0
Black/African American	239,598	13	16.9	241,678	13	17.0	252,566	13	17.1	272,363	13	17.0	256,756	13	17.1
American Indian/Alaska Native	14,217	1	18.0	14,263	1	18.0	14,711	1	17.9	16,183	1	17.7	16,135	1	17.5
White	1,034,712	58	22.2	1,038,435	56	22.3	1,057,803	55	22.4	1,119,398	54	22.2	1,062,439	52	22.4
Hispanic/Latino	259,741	14	18.8	281,216	15	18.8	299,920	16	18.9	337,280	16	18.7	347,906	17	18.9
Asian	71,677	4	23.5	80,370	4	23.5	87,499	5	23.9	93,493	4	24.0	96,097	5	24.3
Native Hawaiian/Other Pacific Islander	4,772	0	19.5	5,676	0	18.6	6090.0	0	18.8	6,797	0	18.6	6,503	0	18.4
Two or more races	64,221	4	21.1	70,013	4	21.2	76,066	4	21.2	85,494	4	21.0	86,119.00	4	21.2
Prefer not to respond/No response	110,305	6	20.7	114,136	6	20.7	129,781	7	20.6	159,334	8	20.1	158,083	8	20.3

Source: ACT Testing Report: Profile Report—National. (2017 Graduating Class). Retrieved from: http://www.act.org/content/dam/act/unsecured/documents/cccr2017/P_99_999999_N_S_N00_ACT-GCPR_National.pdf.

influencing how we see ourselves. For example, the high school female may avoid the upper level chemistry class populated by males because she does not see it as appropriate for her to take such a class, and as the young woman continues to succumb to internal pressure to perform her perceived societal role, she may limit her options for higher education and careers.

Research on the impact of stereotype threat, including Steele and Aronson's (1995) study, suggests it negatively affects career aspirations and performance. In their desire to explore whether stereotype threat applies even when females are interested in STEM disciples, especially math, Shapiro and Williams (2012) found that "the transfer of gender-related math attitudes to girls can put them at risk for self-as-source stereotype threats, stereotype threats rooted in the concern that a performance could confirm in one's own mind that the stereotype is indeed true of oneself or the group" (p. 13). In other words, when women and girls believe that others—society, teachers, mentors, or parents—hold stereotypical beliefs that females are somehow less able to compete in STEM disciplines, such as math, those same women and girls succumb to the fear of proving the negative stereotype about not only themselves but also of other females.

Lest we think stereotype threat applies only to women, it doesn't. No group fully escapes this hazard. Aronson et al. (1999) found that high-achieving white males, as determined by ACT scores, under-performed on challenging math problems when they were subtly reminded that Asians generally performed better in math assessments. Moreover, the long-running stereotype that seems to operate in tandem with the "girls don't do math and science" belief is that "boys don't read or write well." Pansu et al. (2016) tested 80 third-graders, 48 of whom were boys, and found that when they staged the experiment as a "reading test," the boys' scores were in keeping with the stereotype, meaning they scored, as a whole, lower than the girls did. When the experiment was staged as a game, however, the boys outscored the girls. This study indicates the negative impact of ste-reotype on boys, and importantly, it impacts them early. While the authors do not draw conclusions beyond the empirical implications of stereotype threat on boys' reading scores, I think we can see that girls benefit from the perception that they are better at reading *unless* the assessment is framed as a game, which boys stereotypically excel at. Neither group truly benefits from such stereotypes and the accompanying internalized threats. Moreover, given the importance of literacy throughout education, regardless of chosen discipline, categorizing one group as innately less able to read may impede that group from engaging academically and instead encouraging them to drop out. Disparaging one group's reading ability is no better than disparaging another group's math ability. To be clear, it is never one choice, but rather it is the piling up of choices that seem insignificant yet have lasting effects. As the internalization of societal beliefs, stereotype threat is subtle and convincing. Numerous women admitted to fears that people—supervisors, colleagues, and even friends—would think they were hired because they were female and thus needed to fill quotas rather than because they were competent engineers, researchers, or computer scientists.

ANNE'S STORY

*You know, in the early 1980s, things were changing as far as social practice. At least for some people. Not for my mother, however, who held very traditional beliefs of a woman's proper place. She believed women were to get married and raise children. Careers were for men so they could provide for their families. My father was a good provider: as an engineer, he always earned enough that my mother could stay home and focus on family and home. My dad told me I could be an engineer because I was good at math and that I could make good money **until I was ready to stay home**. He thought I could be an engineer, but **he assumed it would be a temporary thing**. I have always felt that pressure from my mother to be married, and I was goal-driven to be married and have kids one day. **I was not really driven toward a career, and in retrospect, I think it's because I didn't think I supposed to be.***

JESSICA'S STORY

I enjoyed my math in high school, but I was definitely afraid to let people see how much of a nerd I was because I worried I would be seen as not pretty or as not willing to go on a date with a guy. I thought the guys might feel threatened if they knew I was the one throwing the bell curve … you know …, back in the 90s, **no one expected it was a girl getting 100 plus extra credit on the tests in math and science classes.**

My high school had a great college prep adviser, and she gave us all sorts of interest assessments to help us figure out where to go to college and what to do in college, what to major in. My interest test told me that I would be a great electrical engineer. So off I went to the University to become an electrical engineer. I had no idea what that meant; I didn't know any electrical engineers … for that matter the only engineers of any kind that I knew were my friends' fathers. **I wasn't even sure if there were any female engineers.** *But I had already decided that I could do it.*

At least until I took my first class in circuit boards. I could do it … but I HATED it. I suddenly felt lost and had no idea what to do.

Fortunately, I had a supportive adviser who worked through this process with me, suggesting that I consider another engineering field such as civil, mechanical, chemical, or biomedical. He warned me, however, that I would need to be ready because I would not see many "young ladies," especially in mechanical. He suggested that I try chemical or biomedical where I would feel more comfortable …

Well, hearing that mechanical was the field with the fewest women, I became a lot more interested in it. I started looking more at mechanical engineering and found it incredibly interesting and to top it off HE DIDN'T THINK I COULD DO IT!

I found mechanical engineering, but it was purely through accidental circumstance and more than a little stubbornness.

Stubbornness was a recurring motif in interviewing women in STEM fields. One woman noted, "Sometimes I am just stubborn and will sometimes stick with it rather than quit." Another stated, "I'm just too stubborn to quit regardless of who wants me to." Stubbornness was generally confessed as a bittersweet character flaw rather than a valuable attribute. Women are not supposed to be stubborn because it is an unattractive attribute despite how it benefits their careers. Interestingly, of the women I interviewed, no one conveyed open defiance to faculty, advisers, or supervisors who questioned their fit for particular positions or STEM fields. Instead, the women reported solitary commitment: "I knew then that I would never quit. I wouldn't give him the satisfaction." Thus, those who had positional authority were not challenged in their stereotypical thinking and were likely confirmed by the women's choice not to confront even outright bias.

Yet these women had little choice; they were in vulnerable positions, perhaps feeling threatened by their own stereotypes, even wondering whether their chosen paths were acceptable paths for women, especially women who wanted work–life balance, some even desiring motherhood.

Targets of prejudicial speech and behavior experience a range of feelings: anger, frustration, self-doubt, confusion, embarrassment, shame, and fear, to name a few. If the targets confront those misguided beliefs, they may be further insulted or worse if their attacker wields power over their academic or professional careers. Those who firmly hold biased beliefs tend to respond defensively when challenged, especially when their targets proffer those challenges. We've all heard those stories from universities and workplaces that rival those of middle-school bullies, except the stakes are higher even as the bias and threats may be more subtle. Lest we assume that only men discourage women from pursuing STEM careers, we must recognize that belief as a myth.

GRETA'S STORY

Lots of young women feel pressure to delay having children or to not have them at all. People think you are more professional and productive if you are at work for longer hours. It's hard but it's incredibly important to keep a work–life balance despite the criticism you hear and the challenges tossed at you.

In the sciences, postdoc appointments are the norm and they are important. I was roughly 3 months pregnant when I started my postdoc, and my primary adviser, a woman, was supportive during the first delivery, but her support evaporated when I announced my second pregnancy.

With my first pregnancy, she told me that if I needed a third year for research, she could cover it with her grant. The week after I announced by second pregnancy, however, she told me she couldn't support a third year for me because she couldn't afford to support someone on maternity leave. Then she hired a male postdoc to replace me. Maybe it wasn't a gender-driven decision, but it sure felt like it.

It was shocking to me because she had children herself. I mean they were grown, but she had confided that she had not felt supported when she had her children …

Greta's supervisor may have questioned her commitment to her career because she had a second child. Young women in many fields are discouraged from having children, and if they choose to have children, the perception seems to be that career-minded women should stop at one child to avoid too many disruptions. Managers and coworkers seem to believe that women who have children are not committed to their careers. Most women commented that managers and colleagues, both male and female, frequently suggested that young women would "probably just get pregnant and leave." People assume that all women want to have children, and that when they have children, the women will likely quit their jobs or at least move to part-time positions. Upon returning to work following childbirth, several young women noted hearing comments like, "Man, I'm surprised you came back to work. My wife just loved our baby too much to go back to work." Most were quick to add that "He meant well, but it didn't come across that way."

JESSICA'S STORY

About 5 years into my career, I got pregnant with my first and only child. I was asked who was going to do my job for me because they assumed I couldn't continue to work while pregnant. Granted this was 15 years ago, but I don't see that much has changed.

At about 5 months, I ended up on bed rest, and I was terrified that I would lose my job. Somehow, I managed to keep it.

When I returned to work, the field engineer asked me who was raising my daughter and told me he couldn't believe I was back at work. I was made very aware that when a woman has a family and continues to work, it's viewed differently than a man who's working. He deserves opportunities to support his family, and when he has another child, people look to support him. He's creating a deeper foundation than a woman who is choosing to do the same things.

Those I interviewed acknowledged that their colleagues generally "respected" their pregnancies, but the pregnancies brought additional stereotypes to the workplace and made gender more obvious. Some men were distinctly uncomfortable around the women as their pregnancies progressed to the third trimester, and this discomfort seems particularly palpable in the story of one civil engineer who works in construction as the only woman in her group.

KAREN'S STORY

People were just very uncomfortable. Maybe it's different in other parts of the country, but I live and work in the South. If I was on site and carrying a clipboard, someone would come up to carry it for me because it wasn't good for the baby.

I had to go out to the field to supervise, and when I was about 4– 5 months pregnant, my boss asked if I wanted to send someone else so the baby didn't get shook out early. Seriously ... he said that and he meant it. He wanted to make sure that I didn't risk harming the baby and apparently riding in a truck on the build site might not be good for me.

I had to explain that it was okay to ride in a truck.

They all thought they were doing things in my best interest, but it just made the situation very awkward for all of us. I didn't realize how uncomfortable many men are around pregnant women.

Other women shared similar stories about pregnancy in STEM workplaces. Most indicated that men tried to be supportive, but they didn't seem to have the interpersonal dexterity to do so: "One male manager [an engineer] asked me how my gestation was coming along. I simply responded, 'Best I can.'"

Breastfeeding created more discomfort in the workplace. For women, breastfeeding was a commitment to their babies, but it brought greater attention to their gender and created awkward and embarrassing moments for them and their coworkers.

KATIE'S STORY

Whenever you have to excuse yourself or mention anything about having to go pump, everyone gets a little embarrassed or something. It's just awkward.

In construction, I tried to be very discreet. They put a deadbolt on my office door and asked if I needed a recliner. I got a separate refrigerator for my breast milk because others were uncomfortable with breast milk there. They were afraid it might drip on their food or someone might have a thirst emergency and drink the breast milk. So, I was given ... but did not ask for ... a private fridge in my office.

TANYA'S STORY

It was hard to find a safe space to pump. People get so focused on their tasks, not so much intentional, but they would interrupt me and walk in on me. It's awkward to plan around the lab's facilities. I mean, I didn't expect it to be a big deal in a research lab. Good grief! We all had to take basic biology and human anatomy.

Yet, other women noted that not having children also leads to workplace expectations: "I don't have children and so I constantly feel pressured to be there, to be at work, for holidays and spring break and perhaps to stay later simply because I'm childless." What motivates the pressure this young woman feels? Perhaps her colleagues assume she simply has greater flexibility in her nonwork schedule because she does not have children, but if we interrogate that more fully, we see both stereotypes and stereotype threat at work. First, "because she is childless," colleagues and her supervisor imply that her life is less busy, less full; beneath this belief lies the stereotypical implication that her life outside of work is empty because she has not given birth, thereby fulfilling some preordained expectation that accompanies her biological sex. She too is at risk of accepting this belief, taking on the extra responsibilities and time constraints because of some presumed failure to uphold her gendered societal role.

Young women often expect mentoring and support from older women in their respective workplaces, but they frequently experience criticism and resentment, as Greta's story conveys. Such behavior does not always come from superiors, as Juanita explains, and gender stereotypes become more complex with ethnic and racial layering.

JUANITA'S STORY

My nickname is the "Queener" because I have a reputation for getting projects back on track and making them successful. Dealing with people who have some preconceived notions of what I can and cannot do is a daily challenge, and I'm never sure how much those notions are driven by my being female and how much they are driven by my being black and southern.

Probably the most recent situation or rather situations involved another woman who you would think would have some solidarity with me since we've worked on successful projects together, but she didn't and doesn't. She is a peer manager in another department, and so we had numerous meetings for a particular project. Most meetings were teleconferences with participants from a variety of locations, including overseas. As I would present information to the group, this woman would publicly challenge me. In every teleconference, she would challenge me, not just question things but challenge me personally. As she became overly aggressive in successive meetings, I simply stopped responding to her during the meetings. I quickly recognized her behavior for what it was: she wanted to upstage me.

The upstaging doesn't just happen from other women. Several women reported feeling invisible to male colleagues, and women of color were particularly sensitive to going unseen by white male colleagues. Most of the women interviewed reported being excluded from invitation-only meetings regarding projects they were working on and in several cases leading part of the project. They report going to meetings and conferences where business cards are shared among male colleagues, and they recall not being included in introductions.

JESSICA'S STORY

The first time I went to my discipline's conference, I was so excited to meet some of the leading researchers and learn more about their more recent projects. Heck, I was just excited to be around so many people who are heavily invested

in research in electrical engineering that you would think I was going on an
extravagant vacation!

Imagine the let down when I manage to finagle my way into a small group
discussion with my main hero of the field only to be excluded from the
conversation—he literally turned his back to me to talk with some young male
grad student. Mind you, I already had a PhD and had cited his work in my
own research. I don't know what he thought, but it felt like he snubbed me
because he didn't think a woman could actually converse intelligently about his
work.

Importantly, the women also convey stereotypical thinking about others in their stories. Jessica, for example, communicates a low appreciation of "nerds" and assumes that males will reject intelligent females, at least ones who demonstrate superior knowledge of math and science. Both Greta and Juanita express shock that women within their fields did not support them *because* they were women; vestiges of "mean girl" stereotypes infuse the feelings of rejection and hurt over other professional women "upstaging," "backstabbing," and not providing a higher level of support than a male might offer.

Yet, we can learn to resist the subtle, and even the not so subtle, pull of stereotypes. For all the stories where stereotypes and biased thoughts intrude to discourage women from STEM fields, I also heard stories of how some mentors, both male and female, encouraged women to pursue such careers. Several women shared how fathers, their own or a friend's, pulled them aside to say, "Hey, you are not going to find the job and make the money you need to make in [that program]. You need to come over to [engineering, computer science, information science, etc.], and I know you can do it." Mentors, whether male or female, can acknowledge some circumstances such as pregnancy that are unique to women while resisting the myth that mothers don't work in careers.

Mentors and peers can also challenge biased beliefs in the classroom and the workplace. Research on confrontation, while incomplete, suggests that witnesses can more effectively challenge sexist beliefs than targets can. For example, Boysen (2013) found that in the college classroom, at least, students want faculty to confront sexism and that in doing so, faculty can reduce the level of sexism in their students. Peers who confront sexist comments garner greater respect than those who utter such comments or ignore such comments. While research is inconclusive on what happens in the face of power differentials, we do know that sexist beliefs will not change unless confronted.

While in graduate school, a peer and I were presenting for the first time at an international conference on communication. As we spoke, one established male scholar began to grill us about tangential issues to the point that even the audience grew uncomfortable. Our mentor, an established female scholar in the discipline, jumped into the fray, quickly conveying that the male scholar should perhaps address the question to her. She confronted, politely but firmly, and then proceeded to thank us for the presentation and essentially closing the session. Over lunch, we discussed the fury of his questions and learned that he was actually targeting our mentor; we were simply fodder in an ongoing power struggle. I've wondered, however, whether he would have attacked us if we were male rather than female.

According to the US Department of Labor's (2016) report, for 61.1% of families with children under 18, both parents work, and 70.5% of mothers with children under 18 are employed (Table 2.2). Moreover, 64.7% of mothers with children under 6 continue to work outside the home. Women work, pursue careers, and continue to raise children. Importantly, women are the primary breadwinner for 40% of the households with children under 18. The halcyon days of Ozzie and Harriet, if they ever existed, clearly do no longer.

Perhaps, Bette L. Bottoms, psychology professor and dean emerita, University of Illinois at Chicago provides sound advice for women who choose to pursue their passions and expand their roles:

"Prepare to be treated as the living incarnation of stereotypes including 'administrator,' 'woman,' and probably worse, 'woman administrator'—but probably not 'leader,' which is reserved for men. Those stereotypes underlie the double standard: Leader-like qualities are praised in men because they are stereotype-consistent, yet the

Table 2.2 US Department of Labor: Employment status of the population by sex, marital status, and presence and age of own children under 18, 2015–16 annual averages [Numbers in Thousands]

Characteristic	2015			2016		
	Total	Men	Women	Total	Men	Women
With Own Children Under 18 years						
Civilian noninstitutional population	65,564	29,095	36,469	65,055	28,992	36,063
Civilian labor force	52,476	26,978	25,498	52,321	26,902	25,419
Participation rate	80	92.7	69.9	80.4	92.8	70.5
Employed	50,238	26,079	24,159	50,240	26,039	24,201
Full-time workers[a]	43,250	24,880	18,370	43,352	24,896	18,456
Part-time workers[b]	6,989	1,199	5,790	6,887	1,143	5,744
Employment–population ratio	76.6	89.6	66.2	77.2	89.8	67.1
Unemployed	2,238	899	1,339	2,082	864	1,218
Unemployment rate	4.3	3.3	5.3	4	3.2	4.8
Married, Spouse Present[c]						
Civilian noninstitutional population	49,822	25,122	24,700	49,472	25,007	24,465
Civilian labor force	40,226	23,532	16,694	40,016	23,409	16,607
Participation rate	80.7	93.7	67.6	80.9	93.6	67.9
Employed	39,026	22,889	16,137	38,866	22,791	16,075
Full-time workers[a]	34,148	21,958	12,190	34,112	21,899	12,214
Part-time workers[b]	4,877	931	3,947	4,754	892	3,861
Employment–population ratio	78.3	91.1	65.3	78.6	91.1	65.7
Unemployed	1,200	643	557	1,150	618	532
Unemployment rate	3	2.7	3.3	2.9	2.6	3.2

Continued

Table 2.2 US Department of Labor: Employment status of the population by sex, marital status, and presence and age of own children under 18, 2015–16 annual averages [Numbers in Thousands]—cont'd

Characteristic	2015			2016		
	Total	Men	Women	Total	Men	Women
Other Marital Status[d]						
Civilian noninstitutional population	15,742	3,973	11,769	15,583	3,985	11,598
Civilian labor force	12,250	3,446	8,804	12,305	3,494	8,811
Participation rate	77.8	86.7	74.8	79	87.7	76
Employed	11,213	3,190	8,022	11,374	3,248	8,125
Full-time workers[a]	9,101	2,922	6,179	9,240	2,997	6,243
Part-time workers[b]	2,111	268	1,843	2,134	251	1,883
Employment-population ratio	71.2	80.3	68.2	73	81.5	70.1
Unemployed	1,038	256	782	931	246	686
Unemployment rate	8.5	7.4	8.9	7.6	7	7.8

Own children include sons, daughters, step-children, and adopted children. Not included are nieces, nephews, grandchildren, and other related and unrelated children. Detail may not sum to totals due to rounding. Updated population controls are introduced annually with the release of January data.

[a]Usually work 35 h or more per week at all jobs.
[b]Usually work less than 35 h per week at all jobs.
[c]Refers to persons in opposite-sex married couples only.
[d]Includes persons who are never married; widowed; divorced; separated; and married, spouse absent; as well as persons in same-sex marriages.

Source: U.S. Bureau of Labor Statistics, 2016. Available: https://www.bls.gov/news.release/famee.t05.htm.

same qualities are held against women, because they are stereotype-inconsistent. A strong male leader is the ideal; a strong woman is a bitch (an aggressively disrespectful term with no male equivalent).[…] Think carefully before you react to unfairness. Do you want to die on that particular hill (confirming the stereotype that women are too quick to take offense)? Or do you want to hold off until you secure a platform from which you can change the underlying system that often condones, dismisses, or fails to recognize sexism?"

THRESHING IT OUT

To follow up on Bottom's questions, how do we pick the hills on which to take a stand? Fool-proof guidelines don't exist because too many variables are at play. We can, however, identify and consider key aspects before engaging:

- Define the hill. What exactly is the obstacle or challenge? Is this Mt. Everest in winter? Or is this hill more like a long, slow climb that's manageable with the right equipment and enough stamina?
- Rate the hill. On a scale of 1−10, with 1 being not important at all and 10 being critical to my career, person, values, or passion, how does this hill rate? How will time affect this hill? Is it already under demolition or is it a permanent structure? How much power is involved with this hill? How important will my success over this hill be in a month, a year, 5 years?
- What are my options for approaching this hill? Might I approach it directly or is an indirect approach better, safer, more likely to succeed?
- What does success look like regarding this hill?
- If you are a mentor, I hope you will imagine how to help women determine the hills worth climbing, and I'm also hoping you're climbing the hills they can't hope to surmount alone. Here's a few suggestions: What are some policies or even just practices (aka "how we've always done things") that seem to affect women more negatively? How might you change those policies and practices?

- Who in your organization presents barriers to women either through their personalities or prejudices? What are some ways you can assist women in working with or around this individual? How might you coax this individual to interrogate his or her beliefs and behaviors?

CHAPTER 3

My Team's Better Than Your Team: How Inclusive Is Office Small Talk?

Contents

Susan's Story	42
Amy's Story	43
Jeanna's Story, Information Sciences Graduate Student	45
Jessica's Story	45
Cara's Story	49
Amy's Story	49
Informants' Comments on Small Talk	50
Amy's Story	52
Threshing It Out	56

> *Football. Hunting. Fishing. Golf. I don't get included in small talk.*
> *~ Emily, engineer*

If you've ever sat with people talking and laughing about the weekend dinner party to which you were not invited, then you know what it feels like to be excluded from a conversation. You don't understand the jokes because you were not at the party; you can't contribute because you lack experience and context. No doubt, you feel a bit awkward for having intruded on such a private conversation, and if you're like me, you are rapidly thinking through escape options. Women in many STEM fields report similar feelings toward office small talk. One informant described it this way: "It's like being the new kid at school trying to break in at eighth grade. It's like a fraternity. It's not blatant exclusion, but it's not inclusive. You are an outsider."

Small talk, what some linguists call phatic communication, acts to unify individuals by establishing rapport, serving to demark the *in*

Communicating as Women in STEM
ISBN 978-0-12-802579-6
https://doi.org/10.1016/B978-0-12-802579-6.00003-3
41

group from the *out* group, *us* from *them*. The stories we share, our humor, even simple greetings, and word choices work to unify us as members of a shared community, and as Holmes and Stubbe (2015) assert, "balancing the demands of the organization's transactional goals alongside the construction of social rapport with fellow workers" is critical for workplace relationships and job satisfaction (p. 165). Sharing personal stories and experiences can work to bond co-workers, demonstrating trust and shared values while building professional intimacy.

SUSAN'S STORY

One of the guys who works directly across from me is unhappily married, moody ... he would just always say things like "I'd be gone already if it weren't for the kids." And it was uncomfortable because some of the things his wife was asking weren't unreasonable. The other guy whose cubicle faced him had already divorced and they sort of bonded over their common misery. It was awkward. Too personal.

Susan conveys how two male colleagues bonded over divorce, but she also shares how she responded differently, feeling repulsed by rather than drawn into the colleague's story. She empathized more with the ex-wife than she did with her colleague and even questioned how much he judged her by the standards he used against his ex-wife. She distrusted him and was grateful when she was reassigned to a different group and no longer had to work directly with either colleague.

Despite the moniker, small talk is complex, serving many purposes at once: "In every social encounter participants are unavoidably involved in constructing, maintaining, or modifying the interpersonal relationship between themselves and their addressee(s) .... [S]ocial talk, including small talk, cannot be dismissed as a peripheral, marginal or minor, discourse mode" (Holmes and Stubbe, p. 89). Small talk isn't relegated to personal social space; it happens frequently in the course of a work day to "break the ice" in meetings or to initiate work-related conversations, much as we usually begin conversations and letters with greetings. When I call or visit a colleague to ask a work-related favor, I generally begin with, "Hey, how are you doing today? Making good progress toward that conference presentation [or whatever we spoke of last]?" If I know the colleague has been dealing with a more personal issue, I may ask about that, something like, "How is your mom doing since her surgery last week?" Only after reconnecting with my colleague on a personal level, do I proceed with the real reason for the contact.

Negotiating small talk challenges newcomers, especially those whose social identities and experiences do not resonate with either the group's current or traditional membership because each workplace and each person in that workplace engages such talk differently. Successful newcomers learn to adapt their personal styles to fit the environment rather than relying on what worked for them in the past.

Small talk resounds with storytelling. Annette Simmons, author of the best-selling *Whoever Tells the Best Story Wins*, explains a key advantage of stories: they "help people feel acknowledged, connected, and less alone" (2007, p. 3). She goes on to explain that effective stories share enough passion and detail to allow listeners to "experience" the feelings and physicality of the storyteller. Storytelling, for Simmons, is both personal and strategic. Yet, research informs us that males and females tell different stories. Overall, men tend to tell stories of competitions, adventures, and activities that promote strength, independence, and success. Women, however, frequently tell self-deprecating stories, even if it is a story of success, and they share stories that focus on people and relationships. Storytelling is gendered, and thus, they "offer a means of constructing particular kinds of communities of practice and workplace relationships, collegial or competitive, self-promoting or other-oriented, supportive or aggressive, and culturally coded as relatively masculine or feminine" (Holmes, 2006, p. 174).

AMY'S STORY

For all my fears of still having a place on the project, I am forever thankful for my colleague and friend, a working mother, who intentionally came by my office on my first day back at work—the first day I had to drop my baby off at day care. She came by just to offer support, to let me know she understands my conflict of wanting to pursue my career while also wanting to be a good mom. She passed along to me the advice she received from another woman: it's possible—not easy, but possible—to be a good mother and also have a successful career. Life is all about balance and boundaries.

This female colleague understood the cultural conundrum of every working mother and affirmed Amy's decision to continue in her career as she learned to define her role as a mother. While many women have no choice in returning to work after childbirth, Amy, like many women in STEM, could have remained at home to raise her child. Her spouse could have supported the young family financially, but she was interested in her career and felt she could contribute to the intellectual aspects of the project. In choosing to return to work, she felt internal pressure to be the perfect mother and the perfect colleague; she also felt the external questioning of why she was at work rather than home with her infant. The mentor's encouragement to find balance and establish boundaries spoke to the internal as well as the external conflicts. Amy's internal conflict and external challenge resonates with other mothers, but it is imponderable for fathers.

Effective small talk and stories entail emotional intelligence (EQ) and facilitate relational practice. EQ is highly prized in the workplace, and numerous entrepreneurs have established themselves as EQ coaches. Writing for Forbes' blog, Caroline Stokes, espouses the value of EQ for leadership and laments, "People often confuse EQ with people-pleasing tactics to avoid conflict, but that's not what it is in practice." We generally define EQ as the ability to perceive our own emotions as well as other people's emotions, to name those emotions in order to use them strategically, and to regulate both personal emotions as well as to help others regulate theirs. While EQ can help individuals manage conflict, avoiding conflict is neither possible nor always desirable. Working toward common goals does not always entail working in unison. Conflict allows us to circumvent group think, which involves limited flexibility and analysis; instead, conflict surfaces when individuals have shared interests or investments in projects and relationships but have different perspectives and different approaches for analysis. EQ offers hope to handle feelings so that conflicts can be addressed productively.

JEANNA'S STORY, INFORMATION SCIENCES GRADUATE STUDENT

Last semester, I was part of a team that had both grads and undergrads. I was the only female. The group of guys were not open to ... you know ... like ... women wanting to better themselves, be professional. Let's say there were old school.

Anyway, we had a breakfast meeting to work on our project. It was just ... awkward ... you know ... to have to sit there, basically ignored, while they ranted about another girl they had on a previous team—a girl they knew was my friend—and they only spoke about her negatively ....

Maybe I should have spoken up, but then they stopped talking about my friend and started talking about drinking, taking advantage of girls, using them for sex, and who all had slept with certain girls as if it were some sort of rite of passage ... and I AM RIGHT THERE AT THE TABLE.

JESSICA'S STORY

At a conference last week, at lunch conversation, somehow the men—I was the only female at the table, of course—began discussing how difficult it is to raise daughters.

One man said he had two daughters and he hoped he wouldn't be cursed with a third.

This comment did surprise me; I'm used to men holding negative conversations about their wives or their ex-wives or their girlfriends, but they don't usually slam their daughters this way. They said a lot of pretty awful things about daughters without thinking once about insulting me. I mean, clearly I am somebody's daughter.

Both Jeanna and Jessica tell stories of men failing at small talk presumably because they either lacked EQ or were mean-spirited. Perhaps the women blended so well into the environment that the men were trying to include them in their average informal conversation; if that was their intention, then it didn't work. When taken to extreme, this sort of conversation crosses over to create a toxic and *hostile* work environment. Were they bullies? When Jeanna told me her story, I thought so. Yet neither woman engaged the conflict. Perhaps both were wisely exhibiting EQ and choosing their battles. The male colleagues in Jeanna's and Jessica's stories did little to build unity or demonstrate camaraderie toward the women present. Rather the competitive and highly critical content that may have appealed to some other men insulted the women.

Relational practice is what we *do*, and sometimes fail to do, through language and behavior in interpersonal interactions. Relating to others effectively requires EQ, a willingness and ability to consider another's perspective. Holmes (2006) describes relational practice as "people-oriented behaviour which oils interpersonal wheels at work and facilitates the achievement of workplace objectives" (p. 75). The behaviors associated with relational practice, as described by Fletcher (1999), are "idiosyncratic, seemingly unrelated fragments of behavior that might otherwise go unnoticed." Importantly, these behaviors are relationally motivated. Fletcher identified four "types" or motivations for relational practice:

- *Preserving:* Behaviors and talk that privilege the project or shared goal
- *Mutual Empowering:* Behaviors and talk that recognize the abilities and contributions of other group members in moving forward on the project or shared goal
- *Self-Achieving:* Behaviors and talk, including self-talk, that recognize and promote our personal abilities related to the project or shared goal; may also include a selfish focus on promoting ourselves or our vision of the project or shared goal
- *Creating Teams:* Behaviors and talk that contribute toward unifying group members toward the project or shared goal

The locker room scene from the football movie *Remember the Titans (2000)* presents an excellent use of small talk to build team rapport. If you'll recall, the movie retells the integration of two high schools in Virginia, specifically the integration of the football team. The young men have just had a hard workout during the training camp and racial tensions remain high despite some softening in key relationships. A new player from California, who shows up late to practice sporting long blonde hair, doesn't have the same social hangups regarding race that the Virginia boys were holding on to. To play on the team, however, he has to get a haircut. A few of the black players begin to tease him about his haircut and nickname him Sunshine. He quickly acquiesces, saying, "Sunshine, huh. That's cool. I can dig it." He then proceeds to slap hands in camaraderie. The other player is taken aback at first, but then Sunshine says, "Come on, bro', don't leave me hangin'." Everyone chuckles and then the jokesters engage another white player with the first of a series of "your mama" jokes (clip available at https://youtu.be/cshMYPMKtIU).

This locker room humor tests boundaries and through phatic communication builds team spirit, which is necessary for their shared goal—winning games. Jokes about mothers, which might ordinarily lead to fights, become the connecting theme: the willingness to recognize, accept, and even participate in the ritual emphasizes brotherly unity. Joining in the repartee becomes mutually empowering; as one teammate proposes a joke, e.g., pretending to have a sore back from carrying an overweight mama, and the other teammates demonstrate appreciation for the joke through laughter, both the jokester and those laughing are empowered by showing their toughness, their willingness to accept the joke—even about one's own mother—for the sake of the team, thereby preserving the team's unity. It's a measure of emotional and psychological strength. Creative jokesters are rewarded with greater applause and laughter from the group especially when the joke's target is rendered incapable of retorting. Such self-achievement asserts a perverse leadership ability: an attitude of success in plying the joke as well as an ability to take a hit for the good of the team.

While Fletcher addressed all four motivations in her 1999 work, Holmes applied only three: preserving, mutual empowering (which she called "off-record mentoring"), and creating teams. The men in Jeanna's and Jessica's stories seem more focused on self-achieving, on one-upping each other in degrading women, rather than preserving and developing team unity, and yet there are similarities to the locker room banter from *Remember the Titans*. In contrast, women do not use the same type of banter to build rapport. We will follow Fletcher's framework as we focus on how relational practice applies in STEM workplaces, to some extent sharing Fletcher's interest, and classrooms. Importantly, however, we will interrogate how women perceive and respond to these workplace relational practices.

Relational practice involves behaviors and talk that rarely appear on employee review forms, and researchers in general describe it as invisible, as both "unimportant" and "necessary" (Carlson and Crawford 2011). Because the behaviors and talk associated with relational practice are considered more characteristic of women or "women's work," most researchers find them undervalued in workplaces, especially traditionally male workplaces such as engineering offices (e.g., Fletcher 1999). Relational practice happens in small talk, in meetings, and even during off-hours: as one informant shared,

> We once had a team-building exercise that was about clay-shooting. I was about 8 months pregnant, but I had to go because I was told that I would miss out on team bonding. In my experience, these team-building activities are almost always male-dominated.

What motivated her to attend the clay shooting? Relational practice. She went to preserve her role in the team, to participate in the team-building experience, and to maintain her personal career goals in terms of the team's project. She was already experiencing some angst toward her impending maternity leave and persistent questions from her male boss and colleagues about whether she would actually return to the workplace. What motivated the team leader to organize this exercise? He likely wanted to create cohesiveness among his team, to empower each member to contribute ideas to the project, to preserve the team's collective focus, and to promote his leadership within the team. The rest of the team, all males, enjoyed the exercise; the lone female did not. Instead, she felt ostracized, singled out, and discouraged.

In the workplace, specifically STEM workplaces that are male-dominated/masculine, men may use small talk or even team-building activities, intentionally or not, to test women in order to determine how well they will "fit in with the guys." For example, speaking disrespectfully of other women, especially when the only female present is a friend of the criticized woman, may serve to determine whether "she'll play along like one of us" or will get her feelings hurt or feel insulted. Such demands can discourage women from engaging in the types of support opportunities, e.g., women's groups or programs, created specifically to encourage them (Cross & Bagilhole, 2002; Hatmaker, 2013). Such interaction forces women to choose between honoring their gendered selves and attempting to be part of the masculine work group, and even if the women choose to deny their feminine selves, play along with the masculine discourse norms, they are then penalized by the very work group for violating their gendered norms.

CARA'S STORY

Well, I'm a fairly young white female engineer, and ... uh ... I guess ... guys generally find me okay looking ...

But, anyway, I'm the only female on my crew, and we've been working in rural areas for a few months. I have to be mindful of my privacy ... I mean, there are anatomical differences ... I can't just go behind a truck! I have to keep some supplies with me ... toilet paper for starters ...

Anyway ... I guess the guys find my personal values and modesty amusing ... they always comment about my leaving to find privacy ... their so-called jokes get pretty crude sometimes ... which is embarrassing and un-comfortable feeling. But I'm never really sure how to react ... you know ... I don't want to be too sensitive ... it's a tightrope ...

AMY'S STORY

Of course, I remember in the 80s, I would act nonchalant over swear words when older men would chastise others for swearing in my presence, and that was uncomfortable because I didn't want to draw attention to myself ...

INFORMANTS' COMMENTS ON SMALL TALK

I probably follow football more—I actually like football—but now I have to KNOW football.

Our company does an annual clay shoot for a charity. I had to learn to shoot even though I had never held a gun. I even paid for lessons so I wouldn't look like an idiot.

My manager golfs ... a lot ... If you played golf, you got invited. If you didn't play, you got left out. I not only had to invest in playing golf but also I had to get decent at playing golf ...

It gets awkward for me when they start talking about the news ... or rather about how attractive the news anchors are and what they would like to do to them ...

Sometimes when I'm tired of hearing the constant barrage of complaints against his wife, I just will say, as light-heartedly as possible, she probably wouldn't appreciate ...

If women are talking, men will go by and ask what we're conniving or are we plotting a takeover because they feel uncertain.

When there's more than one of us (women), it's sort of like a hen party thing—at least that's how we're perceived.

Execution of relational practice may be more important than motivation is. Relating to others requires a willingness to monitor emotions through observing nonverbals and listening empathically; it requires EQ and openness. This oft-touted preference for managers and employees to exercise EQ in the workplace, frequently through popular notions of servant and transformational leadership, seems, at least superficially, to privilege women's stereotypic strengths. Evidence does not support that hypothesis.

Perceived differences between how women and men relate to and interact with others, not surprisingly, follow gendered expectations or stereotypes. Women are perceived as and expected to be better at nurturing, sensing emotions, reading nonverbals, and responding with empathy and compassion. Men, on the other hand, are assumed to be more rational, more competitive, and more analytical. Damore conveys the stereotypes in his now infamous Google diatribe. For example, he claims that men and women have distinct and innate biological and psychological differences that "explain why we don't see equal representation of women in tech and leadership." His rationale follows stereotypic myths that women "have a stronger interest in people rather than things, relative to men," that women are more agreeable while men are more assertive, which supposedly explains why women "generally have a harder time negotiating salary, asking for raises, speaking up, or leading," and that women struggle to regulate their emotions, exhibiting both higher anxiety and reduced ability to handle stress. Damone's implicit fear is that the push for equality in STEM fields will destabilize masculinity, that the push for equality is a privileging of the feminine. Yet a July 2017 MSN poll revealed 20% of men and 21% of women *prefer* to work with men; only 6% of either gender prefer to work with women.

In reviewing 25 years of research in gender and management research, Broadbridge and Simpson (2011) argue "that masculinized or macho management practices are becoming more prevalent in workplaces" and that such practices emphasize the masculine norm of controlling emotions. They further assert that behaviors and methods, including team management and communication practice, once considered feminine are now restated as masculine. When men exhibit controlled emotional expressions, e.g., teary eyes, they are perceived as strong enough to be vulnerable or as having a momentary lapse in adhering to their masculine roles (Shields, Garner, DiLeone, & Hadley, 2007). Women are told they are too emotional.

A number of studies have found that men have greater flexibility in showing emotions in the workplace and that women are penalized both for straying from perceived gender-appropriate expressions of emotions and also for adhering to expected expressions. For example, men are perceived as being emotionally competent and intelligent when they restrain their emotions, but women's authenticity and integrity are called into question when they exercise restraint. Instead, women are perceived as more emotionally competent and intelligent when they immediately react (Hess, David, & Hareli, 2016). This response is something of a catch-22, however, because it coheres with the stereotypes that women are more emotionally intelligent and also less able to control their emotions. In other words, tears and hysteria are part of the feminine stereotype; neither is valued.

Women are expected to understand co-workers' body language and to interact compassionately, carrying an undue burden of emotional support in the workplace. Men are not expected to carry this load, and thus they are recognized—as is their emotional work—as going above expectations. Women are not rewarded for doing this work: they are penalized for failing to do it. The tightrope is indeed thin and treacherous.

AMY'S STORY

I've been an engineer for more than 30 years. It's interesting that men my age still hold conversations about fishing, football, going to the lake, or sometimes their kids, but now I'm working with a number of younger engineers, most of whom are male, but they talk about more things. They discuss tv shows, diets, workouts, fitness … good places to go for dinner … just more inclusive topics. They have actually made me more aware of how much I've missed those kinds of conversations over the years. That's really true because they seek me out sometimes, as if my opinion matters to them. Perhaps some of them see me as a more motherly figure, but I'm ok with that. They see me.

So, how are women seen in traditionally and predominantly male workplaces? Most of the women I spoke with reported feeling invisible at times, and I wondered how frequently they chose that feeling over other feelings, such as anger, disgust, or frustration. When confronted with small talk that alienated rather than included them, most women learned to shrug it off or to avoid interaction with some colleagues when possible. Like Amy, the women I spoke with convey a freedom that comes from being past 45 but younger than 65; they become valued by colleagues, especially younger ones.

Despite our focus in this chapter on the failure of small talk, I heard stories of successful male—female workplace interactions, and the women reported that indeed most of their interactions were professional. Only sometimes, in certain contexts and with a few individuals, did the interactions sour. Families seemed important topics over which males and females bonded. They discussed their children's academics and activities, compared vacation stories, and bemoaned the general busyness of life. Such conversations assisted in establishing rapport so that work could happen. Some discussed holidays and religious obligations; others shared hobbies, including sports, or discussed movies and television shows. Through small talk, they created work teams to achieve corporate goals.

So how can STEM workplaces become more inclusive of women? STEM workplaces must teach employees—new and old, male and female—why we must engage in mindful small talk and then how to do so. Mindfulness, being intentional about our interpersonal interactions, can help us interact effectively and respectfully. We can attend to who is present as we speak, and we can modify our conversation topics and tone to include others. We can check our speech and our thoughts. If we cannot or will not consider the impact of our words on colleagues, perhaps we can consider how our words influence our colleagues' perceptions of us.

Leaders, male and female, need to intervene in alienating conversations to demonstrate inclusion. So do co-workers. How might the conversation about denigrating daughters have changed if one man had dared to differ, sharing thankfulness and pride in his daughter or even commenting directly to Jessica about how her father must be proud of her and what she is doing? Such intervention and attempt at redirecting the conversation may have caused the other men to think before continuing. Failure to interrupt the denigration of daughters in that conversation is tacit approval.

Research on bullying in workplaces emphasizes the value of co-worker response to witnessing and hearing about difficult work experiences. First, we must acknowledge that workplaces are fraught with conflict and aggression, which sometimes develops into bullying (Cowan, 2011; Lutgen-Sandvik & Sypher, 2009). The 2017 Workplace Bullying Survey (http://www.workplacebullying.org/wbiresearch/wbi-2017-survey/) found that 19% of workers experience bullying; another 19% witness it. Roughly 61 million people face workplace bullying in the United States. Bullies tend to be male

(70%); targets tend to be female (60%). Females are the most frequent targets of both male and female bullies, and the impact of bullying in the workplace extends beyond the lone targets. Co-workers, both male and female, also feel stress from witnessing such interpersonal aggression and from deciding how to respond. Research thus far shows that targets of workplace bullying, much like those on the playground, cannot rely on co-workers for support (Namie & Lutgen-Sandvik, 2010). Co-workers fear reprisal and also seek to justify the bullying as a way of explaining their own exemption from the attacks. In short, co-workers tend to distance themselves from the targets of workplace bullying, and in doing so, they increase the isolation and stress victims feel, much like Jeanna and Jessica describe above.

Neither Jeanna nor Jessica considered themselves the targets of bullies; none of the women I spoke with identified as targets of bullies. Most told stories that conveyed discrimination, exclusion, and sexual innuendo. Women of color were more likely to recognize and classify such instances accordingly; the white women I spoke with, however, seemed adamant that the men were simply "insensitive" or "socially awkward" or just "didn't mean anything by it." Such responses keep within the gender-prescribed norms of avoiding conflict, but they also indicate another norming. The women acknowledged that these men were unaccustomed to working with women: "you know, engineering has just always been a man's field, and they just don't know what to do with us women" (Amy).

In their research on stories of workplace bullying, Tye-Williams and Krone (2015) categorize targets' stories as chaos stories, reports, and quest narratives. Most people in their study of both men and women targets told chaos stories, which the researchers describe as lacking in chronology, as "unfinished narratives about unfinished experiences" (p. 11). None of the women I interviewed shared chaos stories. Instead, they shared report-like stories that focused on key details and factual accuracy, like Jeanna's and Jessica's stories, as well as stories that described the experiences as personally beneficial, as quest narratives, much like Darla shares:

You know, when I was in grad school, I was the only female in the chemistry PhD program. The guys, most of them anyway, just didn't want to work with me. They would form study groups but not include me. They would go out to the bar on Friday nights, but I wasn't invited. And I have to think their resistance was either because I'm black or because I'm female. I've often felt this internal conflict in rationalizing whether issues related more to my being black or female.

Anyway, while in grad school, I didn't have any female professors, but several lab assistants were women who had stopped at the master's degree. They were not always my biggest supporters; sometimes I felt they envied me or perhaps they thought I was too uppity since I didn't settle for a master's. Regardless, I just kept plugging along, doing what I had to do in order to complete assignments and conduct my research. Usually alone.

There were several instances in classes, one professor in particular, where we had to write free responses to questions, and I felt my responses, in comparison to my colleagues, were correct. They got 15/15; I got 10/15. When I approached the professor, he told me that he would find other points to take off.

I had to find some other way to be better. I couldn't let him win. I couldn't let him beat me down and make me settle for less than my goal.

But, you know, those experiences make me a better teacher, a better chemistry teacher. I give objective tests. I tell my students that I make mistakes and they should come to see me if they have any questions about their grades, especially if they think I have erred in grading.

Following an almost perfect quest narrative, Darla describes a solitary journey in which she had to prevail against difficult circumstances. Success entailed a solitary persistence, a stubbornness to fight the good fight. Such experiences, which she does not name as exclusive or discriminatory, improve her character and her ability to be "a better chemistry teacher" and to be more attentive to her students' needs. Her quest ends with a traditionally feminine outcome: she is better able to meet the needs of others. She still focuses on relational practice, on mutual empowerment, preserving, and team goal achievement.

Mentors can offer women strategies for handling awkward situations. Simply getting some warning that such interactions may happen may assist young women in preparing themselves. Young women also need to talk through options for responding to egregious violations of workplace decorum. When and how might the young woman need to challenge the colleague? What are potential repercussions? Sharing our own stories may help others.

Importantly, STEM professionals, male and female alike, need to name exclusionary practice, discrimination, and sexual innuendo or harassment. They need to interrupt such behaviors by derailing conversations, seeking clarification, and interrogating their own communication practices. Importantly, they must consider the gendered implications of small talk topics and seek inclusive topics, such as current events, film and television shows, seasonal topics (e.g., spring planting), new restaurants, and health and fitness, to name a few.

THRESHING IT OUT

Let's think about small talk. In the last few conversations you've had with colleagues, what did you discuss? How inclusive were your topics? Who was around during the conversation? Might a colleague have been insulted or turned off by the language or topic? Why? Were YOU insulted, or did you feel uncomfortable for yourself or another?

What was your response? How might you have changed that situation? How could you have intervened to change the topic? Brainstorm some ways to change the topic or redirect how it's progressing.

How might you handle a situation in which the small talk is offensive or disrespectful of a person or group—even when that person or a person of that group is not present? What are your options if YOU are offended by the conversation?

Women in the workplace must

- Know your options. The EEOC's website is an excellent place to begin: https://www.eeoc.gov/eeoc/publications/fs-sex.cfm.
- Identify mentors and allies, regardless of gender.
- Report sexual harassment.

Mentors and supervisors need to evaluate how they perceive workplace behaviors, particularly regarding relational practice. One place to begin the self-reflection is to identify recent examples when you judged a female colleague as being either too emotional or too cold. Describe the context, the conversation (as you can recall), and the nonverbal responses of other co-workers. Now, imagine a male colleague in that situation. How might your response have differed?

CHAPTER 4

Power, Aggression, and Assertion: Who Gets to Speak?

Contents

Deborah's Story	59
Jade's Story	68
Ann's Story	68
Threshing It Out: Ann's Gems	69
Threshing It Out	70

DEBORAH'S STORY

I think we speak different languages—men and women.

Here's an example: My daughter was about two years old and a tornado was coming and the day care was cancelled. I had been in a meeting and was unaware. My colleagues were all in their offices and knew. No one came to tell me. My baby was the last one at day care.

If I had one female colleague, that wouldn't have happened.

Popular book titles over the past 20-plus years reflect the commonly held belief that men and women simply communicate differently:

- *Men are from Mars, Women are from Venus* (John Gray, 1992)
- *Why Men Don't Listen and Women Can't Read Maps: How We're Different and What to Do About It* (Allan Pease and Barbara Pease, 2001)
- *Act Like a Lady, Think Like a Man* (Steve Harvey, 2009)
- *Women Don't Ask: Negotiation and the Gender Divide* (Linda Babcock & Sara Leschever, 2007)
- *How Women Decide: What's True, What's Not, and What Strategies Spark the Best Choices* (Therese Huston, 2016)

As these book titles indicate, our common belief is that men and women, in general, not only use language differently but also think

Communicating as Women in STEM
ISBN 978-0-12-802579-6
https://doi.org/10.1016/B978-0-12-802579-6.00004-5

differently. Researchers from linguistics, sociology, communication, psychology, and cognitive sciences continue to explore these perceived differences. Deborah Tannen (1990), a linguist who focuses on gender communication, describes men and women as operating with different world views that require different ways of using language. She notes that men participate in the world "as an individual in a hierarchal social order in which [they are] either one-up or one-down." Women, in contrast, perceive themselves as part of a "network of connections" (pp. 24–25). These frames resonate in Jessica's story:

> When I was first promoted to project manager, I began to really see how differently men and women think. Women are multi-taskers. That drives some of my co-workers crazy. I can't jump around from plant to plant. We have to handle them one at a time because the guys can't do it.

> They get confused when I would express concerns because I saw everything as interrelated.

> I could tell when I was losing one guy, and we would start over. I don't mind repeating usually. Initially, I would start with "this is what we're going to do and so we'll need ..." you know ... I would start with the big picture, but the field group didn't want to know this.

> Those guys did care about the big picture but only after they learned what specifically is expected of them. They needed "these four might change." They don't want to hear the why first; they want the what. What do you want me to do? They need to hear that before they can listen to anything else.

Jessica's story reveals several aspects of how we think about gender and communication. First, Jessica claims that she and other women can see the "big picture" and want or perhaps even need to see that larger goal first; in other words, Jessica believes that she is representative of other women and that women, in general, are more "big picture" or process-oriented than men are. Moreover, she claims men want to know what she is asking of them, what expectations they will need to meet. In brief, men are more task or outcome-oriented. These ways of thinking are generally supported by research. Women tend to be people and process-oriented; men tend to be task and outcome-oriented. In comparing how men and women interact, researchers have contrasts that may be better explored as trending positions rather than absolutes, as shown in Fig. 4.1. Let's devote a little space to what these gendered features look like in practice and how they can lead to miscommunication in the workplace.

Communication styles reflect cultural expectations (Burleson, 2003). Earlier research in gender communication identified some key differences between interaction and speech patterns of males and females, and later research has further explained those differences, emphasizing that the differences, while salient, may be less extreme and more context sensitive than originally posited.

Speech reflects whether a speaker privileges autonomy or collaboration. For example, if I value collaboration, I will likely ask more open-ended questions rather than ask closed questions or make assertions:

• What are your thoughts on addressing this problem? (open-ended)
• Do you support option A or option B? (closed)
• Option A offers the best solution to this problem. (assertion)

Open-ended questions invite others to share ideas and contribute to the conversation without precluding the solution. If I view the conversation as a contest that I must win, then I will speak to dominate the conversation, and I may even interrupt other speakers to emphasize my points. I will speak in ways that project confidence (e.g., Of course option A is best because …) and deters opposition (e.g., Option B isn't viable at all).

If I see my role in the conversation to facilitate discussion in order to find workable solutions, then I will ask questions to get various

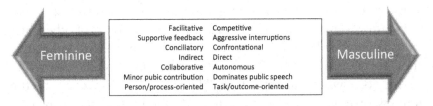

Figure 4.1 Continuum of masculine and feminine interactional styles. *(Based on the list from Holmes (2006, p. 6)).*

points of view, being mindful not to preclude any idea that's put forth but rather to encourage more ideas and further analysis of such ideas. My verbals and nonverbals will convey supportive feedback. I will look directly at speakers to acknowledge them, nod in agreement, and engage in active listening by restating their comments and asking for clarification. To promote further discussion, I will couch criticism in conciliatory and indirect frames, using qualifiers and mitigating aspects to soften disagreements: "Perhaps I missed your explanation of how backflow will be addressed in option A. Would you mind explaining that again?" versus laying down a challenge like "You failed to explain how backflow is handled."

Workplaces that are traditionally filled with one gender tend to perpetuate interaction patterns favored by that gender. Thus, nurses, traditionally a female-dominated field, tend to value more feminine communication styles that are relational, person-focused, and collaborative, while engineers, traditionally a male-dominated field, employ more masculine communication styles that are competitive, confrontational, and task-oriented. Miscommunication sometimes results from these different preferences. For example, nodding one's head as a back channel to encourage speech may be incorrectly viewed as agreement with the speech. Using qualifiers is sometimes inaccurately perceived as lack of confidence in either one's self or one's views. These mistaken assumptions frequently occur in female/male communication.

In each interview, I asked the simple question: How does turn-taking happen in meetings? The overwhelming response: *The loudest voice wins.* This response was not just echoed in one workplace or even in one state, but rather women answered this question across several fields and numerous states. In most cases, the women did not accuse male colleagues of shouting but rather described men speaking very loudly to make their voices heard above the voices of other men and women (See Rogers and Jones (1975) for a primer in dominance strategies in holding the floor). The man who could project the loudest voice would take the floor. "The loudest voice" connotes the quintessential aggressive interruption. Women who tried to project into the din were called "shrill" or worse. So how did these women respond to such grandstanding and silencing?

Some women chose to go to a supervisor or project manager outside of the meeting to express thoughts or ideas individually, but these women frequently saw their ideas quickly usurped by those supervisors with scant credit for their contributions. Others saw this behavior as normal and worked to "earn" the floor through years of service, status through promotion, etc. As one noted, "Eventually, something almost magical happens and you realize you get to speak and are heard!"

I asked a few women why they didn't simply stand up in some of the meetings in order to take the floor, literally, but they found my question outrageous. They could never do that. When I pushed, noting that they had legs and feet that seemed perfectly functional, they categorized such behavior as "inappropriate," "too pushy," "unprofessional," even "rude." Did they view the loudest voice in any of those ways? No. Standing to garner attention need not be done aggressively, per se, but I believe it could reflect assertive posturing.

To substantiate this loudest voice phenomenon, I queried some male engineers and heard that indeed it happens and is perceived as normal. The most interesting story came from a young male engineer with fewer than 5 years of experience:

Well, yeah, that happens. I mean, how else do you get your point across or keep the project from taking a wrong turn? It happened yesterday even.

I was in a meeting about my project and a manager for another group started making ridiculous proposals to challenge my project. Finally, I just started talking over him. He didn't want to stop talking and let me explain or respond so I just had to talk louder than he could. Since I have a loud voice anyway and actually have experience in projecting my voice, he didn't really have a chance.

Given the differences between male and female voices, women have little chance of being the loudest voice. Moreover, projecting over other voices evokes competition and risks confrontation. Women who openly engage such masculine communication styles are judged more harshly than men are (To get an idea of how this works, see Wolfe & Powell (2006), Wright (2016), Tanaka (2015), and Rudman & Fairchild (2004)). Thus, women are put in a double bind: they do not get to speak unless they break gender norms but then are heavily criticized—and penalized—for breaking those norms.

At the risk of being pedantic, I would like to out point a few things. The male engineer did more than proffer the loudest voice: he narrates a very masculine communication style, one that is prized in his male-dominated workplace. First, his speech focuses on the task—protecting his project. He spoke directly in describing the situation; the other speaker had bad ideas and didn't want to yield the floor. Rather than leaving it at that point, the young engineer competes: *"I just had to talk louder than he could."* Thus, he uses his superior skill to claim victory in gaining and keeping the floor.

As noted, women may resist such masculine communication practice, opting instead to be collaborative, adaptive, indirect, and person- or process-oriented. Society encourages such practice from women.

Feminine and masculine communication styles are not absolutes but rather reflect tendencies to interact in certain ways as well as cultural propensities to interpret such interactions in terms of gendered expectations. As Holmes notes, "Gender is always there—a latent, omnipresent, background factor in every communicative encounter, with the potential to move into the foreground at any moment, to creep into our talk in subtle and not-so-subtle ways" (2006, p. 2). As one female graduate student in an operations management program describes it, she stopped noticing how few women were in her program after the first couple of semesters and convinced herself that she and other women are treated equally in the program, in internships, and in the workplace. Yet, she described how golf was a key topic of small talk at her most recent internship and expressed her disinterest in it. This highly competent graduate student repeatedly denies that gender plays any role in her studies or work but shares numerous instances that belie that denial. Several statements throughout our conversation reflect this tension:

More soft-spoken women have more problems.

Smarter people just talk more. The one who knows the most info talks. [She was referring to men.]

One female per team reflects the workplace.

To make it work, girls in this program have to be personable but not distracting. We have to network and be friends. We have to be adaptable. We can't be the outlier.

One girl got kicked out of her team because she was afraid they were missing stuff and so she reached out to teams to make sure her team was doing it right. The guys didn't like that and so they kicked her out.

The negative experiences are just going to happen.

This particular young woman was more direct and halting in her speech, more masculine in her interaction pattern. She responded to my questions but embellished very little. Importantly, she excused the men's behavior by referencing the women's actions. For example, women who are soft-spoken, which is a more feminine trait, experience more interruptions from their male peers than to do women, presumably like this informant, who speak more loudly. Yet she refers to women as "girls," a diminutive term in comparison to the "guys." When asked how team members can be distracting, she referenced how some women would flirt at meetings and/or wear clothing that she found more provocative than her own conservative pants, tops, and jackets. Moreover, her comments display naiveté regarding women's tendency to acquiesce to men's speech, particularly in task-oriented discourse, and men's tendency to interrupt women's speech in these same conversations. As she enters the male-dominated field of operations management, she needs to believe that competence outweighs gender in how she will be received.

Women who have moved into managerial positions and run meetings tend to move away from the Wild West forums to structured discussions that feature an agenda with scheduled times/turns for individuals to speak. As one medical researcher stated, "I knew there had to be a better way to run a meeting so that everyone could talk rather than just one or two egomaniacs." Still another engineering manager explained that she set the agenda and shared it ahead of time so that

everyone knows what we'll be discussing and who will have the floor and for approximately how long you'll have floor. That way, we can leave the meeting without somebody being excluded. If you want to talk, you let me know, and I'll add you to the agenda.

Of course, I guess that means I may be creating a new way to silence people, but so far I think it helps. If somebody really has something to add and they aren't on the schedule, they can raise their hand, and, time allowing, I'll let them talk.

It seems reasonable that leaders in STEM fields can assess their standards of practice in organizing and managing meetings so that everyone has a real opportunity to speak. Perhaps some meetings can allow for open discussion, but leaders must be mindful of who does and who does not speak in those meetings. Similarly, some meetings may require more structure to allow more voices to be heard. While we identify some communication styles with males and others with females, we also know that individuals adapt styles that cohere with their personalities. Thus, some females may speak from a more competitive stance and interrupt aggressively, even doing so mindlessly, and some males may resist such confrontational risks, choosing instead to speak more collaboratively and provide supportive feedback, verbally and nonverbally, rather than interrupt.

Differences in communication styles and expectations translate into miscommunication, and women tend to suffer more from this consequence. Women are chastised for not speaking up; male colleagues presume the women have nothing worthwhile to contribute to the conversation. Since men hold more positions of power within STEM fields, their opinions carry greater weight. Many women feel insulted and slighted by men who talk over them, and if they try to talk over men, they are considered "shrill" or "bossy." Women are expected to show emotion but not too much emotion (Hess, David and Hareli, 2016). Those who fail to show enough emotion are rejected as "cold" or "bitchy." Men who show emotion are considered soft; those who do not are considered an "average guy."

While men may perceive loud "spirited" discussion within a meeting as "just how things get done," women more frequently interpret those discussions as "rude," "like bar fights," "hostile." Women describe feeling that colleagues did not respect them. Only later, after time in their respective fields, did women gain perspective to understand "men arguing doesn't mean disrespect." As one female STEM professor phrased it, "41% of STEM professions at the entry level are women; by the mid to late 30s, however, 50% of those women will leave—not the workforce, just the STEM workforce. They just don't understand what it means to be in the field, and so they internalize the issues."

Women are taught to be indirect, to consider another's perspective rather than framing questions for their own ideas or needs. Thus, women are more likely to say, "Excuse me, I don't mean to interrupt, but ..." or "Would you mind listening to another idea?" than to more directly state "Your idea has merit, but I think ...." Again, men perceive these feminine moves as displays of weakness; women see them as attempts to build community, as simple politeness.

So how do women and men learn to communicate more effectively and inclusively in STEM fields? First, we can take some time to assess the interaction patterns of our workplace. Which styles prevail? Then we can learn to adapt our styles to work strategically within that workplace with an eye toward influencing the workplace to communicate more effectively. Note that I am NOT suggesting that one gender has greater responsibility for analyzing the communication practice of a workplace, but rather I suggest EVERYONE needs to reassess how communication happens, what patterns are privileged, and which patterns are rejected. Importantly, we need to evaluate why those patterns are used and whether they work for everyone.

JADE'S STORY

I was in a meeting that had to have a decision based on data and logic that I had developed. I was the only female. Everyone questioned my data and logic. And we didn't make a decision because they questioned my data, my calculations.

I went to this manager after the meeting to see what I needed to do differently, and he said, "You're just being too aggressive."

Everything he described to me was what I thought was my job. So I asked him, if a man had come in with the same approach would you have thought any differently? He said he wanted some time to think about it.

Later, he came back and admitted he would have considered it strength.

This manager's response is not uncommon. Women routinely receive information to communicate more effectively, perhaps especially at work (see Cameron (1995, 2012), Edwards and Hamilton (2004), among others). Second, we need to assess interaction style for content and context. So, if we find ourselves feeling silenced, we need to figure out why. Are we uncomfortable because colleagues are speaking more loudly than we are accustomed to hearing? Are they angry or simply engaging in verbal combat? How can we engage the conversation and potentially shift it? In other words, we must work to avoid internalizing interaction and see it for what it is: colleagues engaging in verbal competition. We can then strategize our own plan for entering the playing field.

Women entering male-dominated STEM fields need mentors and advocates. Mentors can help women navigate the communicative norms of the workplace even while working to change those norms. Sometimes the problem is NOT miscommunication, but rather it is a refusal to accept either the communication or the person speaking.

ANN'S STORY

It was the worst experience of my life. The group refused to even listen to my presentation. They cut the meeting short and requested a meeting with my supervisor. They told him point blank that they did not want me—a woman—on the project. My supervisor told them they could have me or no one.

While it may have been admirable that he stuck up for me, the experience was terrible. Every day. It was terrible. This one guy would interrupt me and say, "I don't know what you're saying and I don't think anyone else does either. You just need to stop wasting everyone's time."

Someone finally stood up—after a year and a half—and said, "I think she's perfectly clear."

It literally took that long. To make matters worse, he was forced to apologize—which he did reluctantly and without sincerity.

As Ann recounted this story, I was struck by the image of a recalcitrant child coerced into apologizing to a sibling—the empty words "I'm sorry" spoken with venom. When asked what advice she would offer to young women in the field, she offered three gems:

1. Have confidence in your message. You were chosen for your position and are qualified.
2. Nobody deserves to be mistreated. The person doesn't have to like you, but he does have to respect you.
3. Meet with the leads, male or female.

THRESHING IT OUT: ANN'S GEMS

1. How can you demonstrate confidence in your message? Speak up and do so boldly. State your message clearly, directly. Use assertions rather than equivocating questions. If you see a design flaw or significant oversight, point it out, focusing on the content without making it personal: e.g., The current design fails to consider .... OR This approach does not address ....
2. Personal feelings cannot get in the way of the work. Not everyone you work with will like you; you will not like everyone you work with. Did you like everyone you went to school with? No. Did everyone adore you? No. You must, however, give and demand respect. We can listen ... really listen, and we request that colleagues listen to us. We can say things like, I didn't interrupt you, and I expect the same courtesy. It's okay if others think

we are pushy. If we are leading the meeting, we can implement strategies to control interruptions and ensure that everyone is heard. Importantly, those who have status within the workgroup can give up the floor sometimes to those who are trying to be heard. We can even invite them to speak if that's what it takes to give a female colleague opportunity to speak. Sometimes we may have to remind other males to yield the floor for a change.

3. Communicate with those in charge of projects or work groups. Ask: *When I have an idea, concern, or question about an agenda item, how do I get to talk?* Tell: *I'm mainly interested in XYZ, and my goal is ABC.* Consult with a mentor or manager, if appropriate: *In the last meeting, I was not able to voice my concerns. How do you suggest I communicate more effectively in the next meeting? etc.*

THRESHING IT OUT

Keen observations of workplace communication, especially in meetings, can reveal a lot. In your next meeting, note

- Who interrupts? Does everyone have to interrupt—take the floor—or is it only a few who interrupt? Do speakers overtalk everyone or only some people?
- Who gets interrupted? Is it women? Lower-ranked individuals? Is there a pattern?
- How are the interruptions stated exactly? Are they assertions or questions?
- Where do those who get talk sit? Do they sit at the table or toward the front or a particular side of the room? Do you sit there as well? If not, why not?

Mentors must also analyze the communication patterns in the workplace, especially in meetings. If you see imbalances in interruptions, questioning, and turn-taking then you must assist women in participating in the meetings. Perhaps that means changing how you run meetings; perhaps it means directing specific questions to otherwise silenced colleagues.

CHAPTER 5

The Power of the Suit: Dressing for Success According to Whom?

Contents

Threshing It Out   77

> *The higher that women go, we need to dress more professional. We must wear a jacket. If we wear a sweater, we must wear a scarf. Nothing ostentatious.*
>
> ### *Jeannine*

How women dress remains open to public critique, just ask former supermodel and current First Lady Melania Trump who has been criticized for wearing heels to board a plane and for wearing or not wearing particular designers. We may excuse such comments as the cost of being in the public eye, the fate of first ladies and female celebrities. Yet, how women dress arises frequently in typical conversation. Women apologize for being underdressed, for wearing jeans, or for not wearing makeup. Women question whether they look alright or whether their clothing looks good or is appropriate. We spend money and time considering wardrobes. While our daily attire probably doesn't make headlines, it does influence how others think of us and conveys something about what we think of ourselves. Dress came up in most of my interviews even when I didn't initiate the topic.

The idea that we can shape our future success by how we adorn our bodies isn't new. John T. Mollow published *Dress for Success* in 1975 and *The Women's Dress for Success Book* in 1977. The books told women to wear suits, eschew bold prints and colors, stick to blues, blacks, and neutral shades. We were told to avoid excessive jewelry

Communicating as Women in STEM
ISBN 978-0-12-802579-6
https://doi.org/10.1016/B978-0-12-802579-6.00005-7

and sensual or trendy hairstyles. Google an image of Margaret Thatcher or Diane Keaton's role in the 1987 film *Baby Boom* to see the basic model of how women should dress for success. The clothing downplays women's physical bodies, at least in theory. The suits don't work well for every female body, and they can conflict with what some women prefer to wear. Yet, the style, albeit somewhat updated, persists as advice from mentors, both male and female.

Women I spoke with echoed prior research, noting how women must mind their dress because they are "entering a village of men." Young women are told to "dress for your brand" but to avoid "using our bodies to sell ...." Women in STEM need to "wear things and colors that flatter without looking like you're going out for cocktails," "don't wear clothing that is too tight and leaves nothing to the imagination," etc. The power dressing model may prevail as career advice even today, but it does so with significant criticism.

As other researchers have noted, women of color feel particularly trapped by the power suit model. Some women of color have felt backlash for "dressing like white women instead of like myself because I quit wearing big earrings and bright colored shoes." Latinas reported internal conflict between dressing in their preferred culturally driven feminine style and the masculinized style promoted as appropriate for women in STEM (Williams et al., 2016). Homosexuality and transsexuality further complicate this notion of appropriate dress, and little research currently exists on this perspective.

Attire influences how we feel about ourselves. If we feel comfortable and even attractive in our clothing, we are more likely to convey confidence and competence to those around us. Our dress also indicates whether we conform to or rebel against societal norms and how much we do so. Jeannine, quoted in the opening of this chapter, claimed to "understand that view where I'm not going to let anyone change who I am by telling me how to dress" accepts perceived constraints on her dress as a pragmatic expectation of the workplace. The women I spoke with offered up examples of dressing in ways to reduce being confused as administrative assistants or custodians, and I was at times surprised by their urgency in noting the role that attire seemed to communicate professional status. As one engineer explained, "*Well, we* [female engineers] *have to look the part. We can't just dress like the administrative assistants who don't have our degrees and qualifications. I mean … nothing against them, but we have to dress a little classier.*" Style choice is equated with professional status.

Williams et al. (2016) report that women, specifically Latinas, felt the need to present themselves in "culturally neutral" ways which meant a modest use of makeup and clothing with neutral or toned-down colors and clothing that didn't fit too snugly. Women of color felt the need to dress up more than their white counterparts to demonstrate professional status and to reflect social support from their home communities.

Dress becomes more complicated for pregnant women and new mothers whose bodies do not conform to the straight lines of either the power suit or tailored clothing. Determining how to dress, particularly during the last trimester, is challenging, and when women work in environments where they are told how to dress and how not to dress, especially when the prescribed dress does not cohere with their bodies, they can feel extra stress.

Pregnancy gives gender greater presence in the workplace. Men frequently are ill-equipped to handle interactions with pregnant peers because they have little experience, as evidenced by stories in previous chapters. Little, et al. (2015) describe strategies women use to navigate pregnancies, the conflicts they experience, and the mixed results of those strategies. Yet, pregnancy neither impacts a woman's ability to do most jobs nor does it sentence the woman to life at home.

Physical bodies reveal differences, and clothing can emphasize or de-emphasize physical, sexual differences. How women clothe our bodies seems to carry greater meaning than how men clothe theirs. Why?

Perhaps we can write it off to vestiges of patriarchy. Women and women's bodies are objectified. Women's bodies are to be admired or insulted; our bodies—and ourselves—are considered trophies, mysteries, desirable … unless they're not. Then they are considered flawed, unattractive, and aged. Some women have and do use their bodies to gain favor; some women's bodies are harassed and assaulted. Some notions of dress codes evolve as unstated victim blaming.

If you want to be respected in the office, you have to dress respectably. You can't wear tight clothes, short skirts, and hooker heels.

Does the blaming and the judging register with us? Do we see the hypocrisy?

Women are criticized for dressing in ways that are perceived as too sexy, and we are criticized for being too frumpy, for not being feminine enough. Where's the balance?

When boarding the plane in Washington, DC, First Lady Melania Trump wore the finest spiked heels; she deplaned in Texas to tour hurricane-stricken areas wearing sensible tennis shoes. Clearly, she planned ahead and behaved in a way befitting a responsible adult: she dressed herself appropriately for her activities. Her shoe choice made national and international headlines. Her husband's did not. As one woman noted in our conversation, *"Personally, I think we have too many options. I mean we don't just get up and grab a pair of pants and a shirt and slip on our loafers."*

Women do have more style options than men have, and the choices we make influence what others think about us and what we think about ourselves. Research in this area is extensive, but Johnson, Lennon, and Rudd (2014) review significant discoveries regarding the social psychology of personal appearance. Here's what we know.

Regardless of gender, being well-groomed matters. Let's unpack the idea of "being well-groomed." For males and females, being well-groomed entails being clean—clean hair, no body odor, clean and neatly trimmed/filed nails, freshly laundered clothes, polished shoes, etc. For men, it may include a fresh haircut and clean-shaven face, although longer hair and beards, if trimmed and combed, have become increasingly accepted in some contexts. For women, this term includes manicured nails and "modest" or "neutral" makeup. Barbers and hairdressers are frequently featured on social media providing such services to homeless men and women, and the before/after photos of their clients are stunning. Who determines what is "modest" and what is "neutral"? Such concepts are subjective, especially in terms of cultural preferences and norms.

In addition to being clean, professionals are expected to dress in ways appropriate for their workplace. As much as First Lady Trump was criticized for her shoe choice, former First Lady Michelle Obama was disparaged for wearing sleeveless dresses. Apparently, first ladies are to keep their arms covered. Who decides what's appropriate? In workplaces, professionals expect each other to dress in a style worn by colleagues or perhaps to dress a little higher, a little more conservatively. The proverbial "dress for the position you want" resonates loudly and for good reason. Those who imitate the dress of superiors tend to benefit in terms of job assignments and promotions.

For men, such sage advice generally means wearing starched shirts, better quality pants, and polished wingtips. The differences are subtle. For women, those differences can be significant. Hillary Clinton's pantsuits became fodder for comedians and pundits, especially during the 2016 presidential election. The unstated accusation in the pantsuit criticism is that she failed to dress enough like a woman: she didn't wear a dress or a skirt. Professional women may wear pants and even pantsuits, but they are also expected to wear skirts and dresses. For the sake of comparison, consider how Condoleezza Rice and Nancy Pelosi dress.

Women in the workplace must dress professionally, and they must also dress in appropriately feminine ways without dressing provocatively. Yes, women have many choices, but those choices carry risks. When women wear clothing that is more fitted to their bodies, have lower necklines, higher hemlines, etc., men and women focus more on the women's bodies than on what they are saying, and both men and women judge those women more harshly. It's not fair, but it is reality.

It's true: you never get a second chance to make a first impression. We can, however, affect our first and subsequent impressions by how we dress. Professional women must traverse another tightrope in determining what is appropriate, acceptable, sufficiently but not provocatively feminine attire. This list extends to hairstyles, makeup, jewelry, shoes, handbags and brief cases, phone cases … everything.

All of it impacts the women's professional brand. Personal branding is a newer, consumer-driven name for impression management. The idea of personal/professional brand has deep roots, but Tom Peters popularized it in his 1997 article, "The Brand Called You." More recently we call this practice impression management, and professionals who are also women must be mindful of how to affect others' perceptions (Hatmaker, 2013).

Business attire in the United States has become more casual. I have no doubt that some will argue business attire has become too casual. In their survey of business students, Cardon and Okoro (2009) note, "Younger professionals clearly associate authoritativeness and competence with more formal business attire, productivity and trustworthiness with somewhat formal workplace attire, and creativity and friendliness with more casual workplace attire." As we seek to develop more diverse workplaces, workplaces that reflect our culture, we must accommodate by broadening dress codes and expectations that promote productivity and trustworthiness, as well as creativity and inclusion. Some of these changes will happen naturally; some will require pushing.

Women will need to consider that they are "entering a village of men," but the village must also consider its need for competent colleagues—both male and female. Women do not and will not wear the same clothes that men wear. Perhaps we can learn from Red Ants Pants entrepreneur Sarah Calhoun who established a line of work pants for women "that would fit, function and flatter working women" (https://redantspants.com/our-story/meet-calhoun/). Calhoun tired of the work pants dilemma women faced and created a solution. She knew women worked at jobs just like men but needed different but comparable clothing.

The notion of personal branding has received sharp criticism for privileging the masculine, white, middle/upper classes. Lair, Sullivan, and Cheney (2005) provide an overview as well as a critique of personal branding. As these researchers suggest, "Rather than focusing on self-improvement as the means to achievement, personal branding seems to suggest that the road to success is found instead in explicit self-packaging." (p. 308). Dress is critical for the external packaging, but personal branding creates a true conundrum for professional women:

> *In particular, personal branding promotes a feminine surface identity and a masculine internal identity, all the while perpetuating the work/ home dualism. Personal branding encourages women to get ahead at work, work as hard or harder than their male counterparts, and reach for the top but also to look womanly, take care of their external appearance, be there for their children and husbands (if a woman has them—but recognize that if she does, she may not be viewed as a 100% company woman), and routinely act in the caretaker role at work (Lair, Sullivan, and Cheney, p. 328).*

Mentors can share insights about expected dress codes, but young women need space to decide whether those constraints are acceptable. They need space to push the boundaries and eventually redefine those boundaries. Female mentors and role models may be the best suited for this task. The idea of men telling women how to dress for work harkens to a history of repression, exclusion, and abuse.

Women in STEM *may* need some mentoring on dressing for success, but they also need some space to dress as they choose. Workplaces must change to accommodate women's freedom to choose how to dress our bodies. While some positions require certain clothing (e.g., it's unwise to wear synthetic fabrics in labs), most don't. Perhaps we can reconsider the value of expecting women to conform to dress codes that limit personal choice.

THRESHING IT OUT

Mentors and colleagues, regardless of gender, must re-evaluate biases toward attire. Why must women conform to a male village? Does a woman's decision to wear or not to wear bright colors, high heels, and flashy jewelry truly reflect mental acuity and training? Or are we distracted because people—men—in our particular workplaces haven't dressed like that?

What does a successful woman in our STEM field look like? Describe her, including what she is wearing. Is our image based on a real person or on our ideal, our biases?

If we see a woman in our workplace who is dressed differently than our "code," what do we think about her? Why? What does our company's or group's dress code say about attire? Does it apply differently to women and men? How might it be improved to better accommodate professionals regardless of gender?

CHAPTER 6

Communicating as Professionals in STEM: Some Closing Thoughts

Contents

Sadie's Story	80
Mary's Story	82
Jane's Story	82
Lori's Story	83

> *One cannot not communicate.*
> **Paul Watzlawick, psychologist and communication theorist**

The first axiom of Watzlawick's theory of interpersonal communication is that "we cannot not communicate." Perhaps the grammarians among us may balk at the double negative, but if we pause for a moment, we can measure the wisdom of that statement. Whether we talk or remain silent, we are communicating. Even when we try to avoid communicating, we communicate. We communicate unintentionally, sometimes inaccurately, and sometimes deceptively, but we always communicate.

If it is impossible to refrain from communicating, perhaps we should consider ways to communicate more effectively, productively, and inclusively. STEM fields are changing, and that change will speed up as more women enter these male-dominated villages. These changes necessitate broadening the communication practices of STEM fields. Everyone has a role.

In this last chapter, I would like to elaborate on three critical aspects of communication: listening, nonverbal, and verbal. We will discuss these aspects with particular regard to those entering STEM

Communicating as Women in STEM
ISBN 978-0-12-802579-6
https://doi.org/10.1016/B978-0-12-802579-6.00006-9

fields, those already in the fields (especially mentors), and organizations within the fields.

Any attempt at improving communication begins with mindfulness—with intentional focus on the context, the goal or purpose, and the participants. In other words, mindfulness in communication entails a rich awareness of the circumstances, of causes and effects; it requires noticing and presence. To be mindful is to listen—actively and empathically. You're likely familiar with the tenets of active listening:

- Attend to the speaker (listen to the words; don't try to multitask)
- Provide appropriate feedback (head nods, eye contact, etc.)
- Repeat what you heard and ask for verification and clarification (*If I'm hearing correctly, you said … Is that right?*)
- Listen to hear rather than to respond (avoid judging and thinking about how to rebut what the other person is saying)
- Respond if appropriate and after the other person has finished (your response should be honest and on topic)

Empathic listening goes beyond active listening and attends not only to the verbal message but also to the nonverbal messages. It involves attending to underlying emotions, needs, and values. In other words, it means engaging emotional intelligence to better comprehend the message and to make sure the speaker has finished before deciding whether and how to respond. Empathic listening is people-oriented rather than being oriented toward content or task. Let's look at an example.

SADIE'S STORY

Why don't they listen?

I really hate to immediately go to it's because I'm a female … um … but I think it makes it more difficult … because it's harder for your voice to carry, in general, like you have to almost raise your voice and sound angry and then they think, you know, that you're the angry female … that's NOT what you're trying to do …

You almost have to yell with a smile on your face to not sound … you know … that's something I'm still working on …

I was on a conference call just last week ... conference calls are the worst too because it's the loudest voice that wins ...

Anyway ... I would sit there and say now guys ... now guys .... You know, just trying to get in ...

I finally got started and someone tried to interrupt and I just said, 'If you'll let me finish, I will let you speak.'

I don't want to get to this point ... but sometimes you just have to be the angry, interruptive female.

What do you hear in Sadie's story? As we sat in the coffee shop, 27 min into our interview turned conversation, Sadie blurted out this story. She was leaning in, somewhat hunched over her coffee, peering intently at me, hoping I could somehow give her an answer that made sense, one that didn't relate to her gender. In some ways, this book is my attempt at answering her.

Sexism is the easy answer, but it is an incomplete answer. The reported pattern of men not listening, making insensitive and insulting remarks, and excluding women is only partially explained by sexist beliefs of both men and women. The men may have been acting out of sexist belief, consciously or not, but so was Sadie. In the South, "ladies" smile through everything, and Sadie apologetically states she's "still working on it." She is bothered by not being able to handle the situation, to exert her voice, without becoming unlady-like and then finally concedes that sometimes it's necessary "to be the angry interruptive female." In this story, Sadie relates her internal struggle between how she wants to interact and how she is compelled to interact. Her concession is quite different from the young male engineer who gloats over winning the contest.

Sadie values fairness; she wants the opportunity to share her knowledge just as she has listened to others speak on this conference call. She values politeness and turn-taking; she needs to be valued as a member of the team. She looks to share rather than dominate the floor. In 2012, the President's Council of Advisors on Science and Technology reported to then President Barack Obama that "many students, and particularly members of groups underrepresented in

STEM fields, cite an unwelcoming atmosphere from faculty in STEM courses as a reason for their departure" (Engage to Excel, RTP, 2012). Clearly the workplaces often fail to roll out the welcome mat. As I spoke with these women, I gained deep appreciation for their resilience.

Dr. Pamela McCauley Bush, Professor and Director of the Ergonomics Laboratory in the Department of Industrial Engineering and Management Systems at the University of Central Florida, lamented the lack of focus on retention in the workplace. While she applauds the research and changes that are happening in STEM education, she expresses concern that as much as 50% of young women in STEM leave by their mid- to late-30s. Women in STEM frequently feel marginalized, isolated.

Some isolation is self-imposed, but women, especially mothers, don't feel supported.

MARY'S STORY

Companies don't do enough to support young moms. They automatically make the assumption that you're going to quit after you're a mom. Then you start getting pulled off of jobs or not assigned jobs and are told it's because we don't want to overburden you but really it's because we don't want to give you this because you're probably going to quit.

Even when you constantly reassure them that you're not going to quit. It's really frustrating.

JANE'S STORY

I don't want to go out for drinks to do work. I want to go home There are times when I don't want to travel. It's really a stumbling block when I'm not allowed me time to take care of my family.

When it's quitting time, I don't want to stay.

I guess it's just lots of little frustrations that pile up …

LORI'S STORY

Almost all the women I know are high performers and in school were always leaders or held leadership positions in programs. Almost all have a leadership trait.

The typical engineer tends to role. You will be the expert in your field and live in your role as an expert. Most are men. Most of the women are not driven to be the SME but want to be the leader. But the desire for leadership is frustrated by the role of being mom. Men don't seem to have that frustration in being a dad.

At the end of each interview, I asked what advice they could offer other women interested in STEM. I'll highlight their responses here.

Find a mentor, preferably another female, even if it requires going to another department. While the women cautioned that not every female will advocate for other females, they do understand what it means to be a woman in that workplace. A few women encouraged newcomers not to overlook men as mentors because they may be their strongest supporters. No one travels the road to success alone; find a guide to help you climb the mountain.

Perseverance and determination are strong attributes of successful professionals, and women in STEM surely possess these attributes. Yet, they frequently label these traits as stubbornness. Perhaps they do so because they see themselves as breaking gendered norms. Young women who persisted in STEM fields frequently did so with the encouragement of mentors. Find a mentor, be a mentor, much like the women in the iCan movement in Alabama.

Be confident in your message. You completed the coursework just like everyone else. You know what you know, and your voice matters. If you ran the data, you know it better than anyone else. If you did not run the data, you have knowledge, background, experience, and critical thinking to ask pertinent questions.

In the late 1980s and early 1990s, airbags were pushed as critical safety features for all automobiles. In testing these passive safety devices, automakers and the NHTSA used dummies that matched the average male: 5'8", 165 lbs. It should have surprised no one that deployed airbags—designed to blast forward quickly and firmly—could seriously harm smaller individuals, e.g., shorter women (between 4'10" and 5'4") and children. Shorter women and children sustained severe, frequently fatal injuries. Where were the critical voices in the design and implementation of supposedly safe devices? Were they silenced?

Be confident in yourself. Sometimes we can know our level of expertise in a given area and still question our ability to convey that message or do what is necessary to see that idea to fruition. We may not even be able to assert our presence in a meeting and our right to speak in that meeting. Women must be assertive in speaking up and speaking out, in taking the floor. Assertiveness is not aggressiveness. Almost every woman I spoke with told stories of being silenced, of being prevented from stepping forward with worthwhile ideas, and of having those ideas claimed by others. Being bold enough to stand up in a meeting and say, "excuse me, but have we considered ..." is not inappropriate and is not unfeminine. Even though some may judge us more harshly for standing up and speaking out, we must do it anyway. Respect is more important than popularity.

Organizations and teams benefit from encouraging members to speak boldly, expressing confidence in their messages and themselves because such courage enables us to combat group think, the stifling, uncritical agreement to see things as everyone else does.

Name harassment and disrespect. In interviewing these 49 intelligent and strong women, I was amazed that they rarely named the harassment and disrespect they experienced, opting instead to excuse the men (and sometimes women) who failed to see them as equal colleagues, as peers, and who chose to insult them, marginalize them, and disregard them. Yet, I also understand their dilemma. Admitting to such disrespect and harassment carries stigma: culturally, we blame women when bad things happen to them. When women are assaulted, we ask if they were alone, out too late, in a bad neighborhood, or dressed too scantily. Did we "allow" ourselves to be cornered alone in an office with the boss who assaulted us or made unwelcome advances? Didn't we play along and invite such advances? Do we ask these things of men?

It's time to challenge the perpetrators. It's also time to recognize that not all perpetrators are men and that not all victims are women.

Power plays the lead role in demeaning others. Bullies, whether they are on the playground, in social media, or in the workplace, act as they do to assert power, to claim power. They may also act, perhaps even unconsciously, to uphold existing power dynamics, to uphold patriarchy as natural, as customary. This power dynamic is not static, however, and will conflate.

The onus cannot rest on women alone. Those already in STEM fields must step up to mentor and intervene strategically. Mentors and colleagues must support women in STEM. Whether you are a professional mentor or a colleague, you can stand up to workplace bullies and old school thinkers. You will be uncomfortable, but how comfortable will you be living with the memory of what you could have said but didn't? As more stories of sexual harassment and assault are exposed in politics, media, and Hollywood, I am left to wonder how those actions went unchecked for so long. It was common knowledge; neither the victims nor the bystanders spoke up. Surely some of those bystanders were as powerful as the offender; surely someone could have called out the perpetrators for their terrible deeds.

Mentors need to help women in STEM interpret interactions through a less feminine lens. Interruptions are not necessarily rude; speaking out is required. Whatever happens in conference calls does not necessarily carry over into other conversations: it's not personal. Women and men must cross gendered expectations and biases, and mentors can guide this process. Mentors can question gender bias in how we perceive others. For example, a mentor can help Sadie evaluate why she is the "angry, interruptive female" and how that differs from the "angry, interruptive male." The mentor can suggest ways to interrupt before she becomes angry, and the mentor can also question why Sadie seems to excuse the men's interruption and overtalk as normal while hers is rude. Mentors can help women process the double binds they experience, and collectively, they can develop ways to mitigate some of those binds.

STEM workplaces must do more to meet the needs of women. That means, these organizations must develop policies and practices that permit women to fulfill obligations to families. Women continue to do the majority of child care, domestic chores, and tending to the needs of elderly parents. Other workplaces have begun the hard work of mitigating these factors, including creating flexible work schedules that permit parents to get children to day-cares and schools, to work from home at times, such as when children are sick or schools take extra holidays. Job sharing is a viable option for some positions. The Family Leave Act has mandated some policies; it's time to make those policies part of the culture.

Women should not be "unburdened" of assignments for fear they will not return. Such practices work against retention and inclusion of women in STEM workplaces. Moreover, those practices adhere to a nonexistent, and perhaps mythical, notion of family life. Remember, 70.5% of mothers with children under 18 are employed. Women in STEM are more likely to continue working than they are to stop working after childbirth. Rather than making decisions for working mothers, managers must allow the women to choose whether a project or promotion is right for them. Organizations must do a better job of addressing workplace bias against working mothers. Thus, conversations about whether a particular employee can manage the workload with family responsibilities needs to include the employee—only the employee can truly answer the question.

Organizational leaders must identify and revise structures that work against women in the workplace. Team building exercises frequently emerged in my conversations as uncomfortable and isolating events that privileged male colleagues. While males and females can and do participate in clay shooting, holding such an event while one team member is ready to go out on maternity leave is not appropriate. Team building exercises require activities that include everyone; they should not always favor one gender. Leaders must plan team building activities mindfully, knowing the needs and preferences of team members; leaders must have the emotional intelligence to plan team activities that encourage full integration of all members.

Organizational leaders cannot expect women to conform completely to male-dominated workplaces; the workplaces must change not just to allow women access but to allow women to succeed as well.

REFERENCES

Acker, J. (1990). Hierarchies, jobs, bodies: A theory of gendered organizations. *Gender and Society, 4*(2), 139–158.

ACT Testing Report: Profile Report—National. (2017 Graduating Class). Retrieved from http://www.act.org/content/dam/act/unsecured/documents/cccr2017/P_99_999999_N_S_N00_ACT-GCPR_National.pdf.

Aronson, J., Lustina, M. J., Good, C., & Keough, K. (1999). When white men can't do math: Necessary and sufficient factors in stereotype threat. *Journal of Experimental Social Psychology, 35*, 29–46.

Babcock, L., & Leschever, S. (2007). *Women Don't Ask: Negotiation and the Gender Divide.* Princeton, NJ: Princeton University Press.

Blumberg, R. L. (2015). *Eliminating gender bias in textbooks: Pushing for policy reforms that promote gender equity in education.* UN Educational, Scientific and Cultural Organization. EFA Global Monitoring Report. Retrieved from http://unesdoc.unesco.org/images/0023/002324/232452e.pdf.

Broadbridge, A., & Simpson, R. (2011). 25 years on: Reflecting on the past and looking to the future in gender and management research. *British Journal of Management, 22*, 470–483.

Bottoms, B. L. (September 1, 2016). *16 Insights for woman leaders.* Inside Higher Ed. Retrieved from https://www.insidehighered.com/advice/2016/09/01/insights-being-woman-leader-academe-essay.

Boysen, G. A. (2013). Confronting math stereotypes in the classroom: Its effect on female college students' sexism and perceptions of confronters. *Sex Roles, 69*(5–6), 297–307. Retrieved from https://link.springer.com/article/10.1007/s11199-013-0287-y.

Burleson, B. R. (2003). The experience and effects of emotional support: What the study of cultural and gender differences can tell us about close relationships, emotion, and interpersonal communication. *Personal Relationships, 10*, 1–23.

Cameron, D. (1995, 2012). *Verbal hygiene.* Oxon, UK: Routledge.

Cardon, P. W., & Okoro, E. A. (2009). Professional characteristics communicated by formal versus casual workplace attire. *Business Communication Quarterly, 72*(3), 355–360.

Carlson, J. H., & Crawford, M. (2011). Perceptions of relational practices in the workplace. *Gender, Work and Organization, 18*(4), 359–376.

Cowan, R. L. (2011). "Yes, We Have an Anti-bullying Policy But...": HR Professionals' Understanding and Experiences with Workplace Bullying Policy. *Communication Studies, 32*(3), 307–327.

Cross, S., & Bagilhole, B. (2002). Girls' Jobs for the Boys? Men, Masculinity and Non-Traditional Occupations. *Gender, Work & Organization, 9*(2), 204–226.

Damore, J. (2017). *Google's ideological echo chamber: How bias clouds our thinking about diversity and inclusion.* Retrieved from https://www.scribd.com/document/356850149/Googles-Ideological-Echo-Chamber-pdf.

Edwards, R., & Hamilton, M. A. (2004). You need to understand my gender role: An empirical test of Tannen's model of gender and communication. *Sex Roles, 50*(7/8), 491–504.

Fletcher, J. K. (1999). *Disappearing Acts: Gender, Power, and Relational Practice at Work.* MA: MIT Press.

Harvey, S. (2009). *Act Like a Lady, Think Like a Man.* New York: HarperCollins.

Hatmaker, D. M. (2013). Engineering identity: Gender and professional identity negotiation among women engineers. *Gender, Work and Organization, 20*(4), 382–396.

Hess, U., David, S., & Hareli, S. (2016). Emotional restraint is good for men only: The influence of emotional restraint on perceptions of competence. *Emotion, 16*(2), 208–213.

Holmes, J. (2006). *Gendered talk at work: Constructing social identity through workplace interaction.* Malden, MA: Blackwell.

Holmes, J., & Stubbe, M. (2015). *Power and politeness in the workplace.* New York: Routledge.

Huston, T. (2016). *How women decide: What's true, what's not, and what strategies spark the best choices.* New York: Houghton Mifflin Harcourt.

Johnson, K., Lennon, S. J., & Rudd, N. (2014). Dress, body and self: Research in the social psychology of dress. *Fashion and Textiles, 1*(20). Retrieved from http://link.springer.com/article/10.1186/s40691-014-0020-7.

Lair, D. J., Sullivan, K., & Cheney, G. (2005). Marketization and the recasting of the professional self: The rhetoric and ethics of personal branding. *Management Communication Quarterly, 18*(3), 307–343.

Little, L. M., Major, V. S., Hinojosa, A. S., & Nelson, D. L. (2015). Professional image maintenance: How women navigate pregnancy in the workplace. *Academy of Management Journal, 58*(1), 8–37. https://doi.org/10.5465/amj.2013.0599.

Litwin, A. (2014). *New rules for women.* Annapolis, MD: Third Bridge.

Lutgen-Sandvik, P., & Sypher, B. D. (2009). *Destructive Organizational Communication: Processes, Consequences, and Constructive Ways of Organizing.* New York: Routledge.

McConnell-Ginet, S. (2011). *Gender, sexuality, and meaning: Linguistic practice and politics.* New York: Oxford.

Namie, G., & Lutgen-Sandvik, P. (2010). Active and Passive Accomplices: The Communal Character of Workplace Bullying. *International Journal of Communication, 4,* 343–373.

National Science Foundation, National Center for Science and Engineering Statistics. (2015). *Doctorate recipients from U.S. Universities: 2014.* Special Report NSF 16-300. Arlington, VA. Retrieved from http://www.nsf.gov/statistics/2016/nsf16300/.

Olsson, S. (2000). The 'Xena' Paradigm: Women's Narratives of Gender in the Workplace. In J. Holmes (Ed.), *Gendered Speech in Social Context* (pp. 178–191). Wellington, NZ: Victoria University Press.

Pansua, P., Régner, I., Max, S., Colé, P., Nezlek, J. B., & Huguet, P. (2016). A burden for the boys: Evidence of stereotype threat in boys' reading performance. *Journal of Experimental Social Psychology, 65,* 26–30.

Pease, A., & Pease, B. (2001). *Why Men Don't Listen and Women Can't Read Maps: How We're Different and What to Do about It.* New York: Random House/Harmony.

Rogers, W. T., & Jones, S. E. (1975). Effects of dominance tendencies on floor holding and interruption behavior in dyadic interaction. *Human Communication Research, 1*(2), 113–122. Retrieved from https://interruptions.net/literature/Rogers-Human CommRes75.pdf.

Rudman, L. A., & Fairchild, K. (2004). Reactions to counterstereotypic behavior: The role of backlash in cultural stereotype maintenance. *Journal of Personality and Social Psychology, 87*(2), 157–176.

Shapiro, J. R., & Williams, A. M. (2012). The Role of Stereotype Threats in Undermining Girls' and Women's Performance and Interest in STEM Fields. *Sex Roles, 66,* 3–4 [Accessed 03/06/2014] https://link.springer.com/article/10.1007%2Fs11199-011-0051-0.

Shields, S. A., Garner, D. N., DiLeone, B., & Hadley, A. M. (2006). 'Gender and Emotion.'. In J. E. Stets, & J. H. Turner (Eds.), *Handbook of the Sociology of Emotion.* New York: Kluwer.

Simmons, A. (2007). *Whoever tells the best story wins.* New York: AMACON.

Steele, C. M., & Aronson, J. (1995). Stereotype threat and the intellectual test performance of African Americans. *Journal of Personality and Social Psychology, 69*(5), 797–811.

Stix, G. (2012). *'Talking Back: Is American Science in Decline?' Scientific American* [Accessed 6/15/2015] https://blogs.scientificamerican.com/talking-back/is-american-science-in-decline/.

Stokes, C. (2012). *What dorie Clark's latest book can teach coaches about emotional intelligence.* Forbes Community Voice. Retrieved from https://www.forbes.com/sites/forbescoachescouncil/2017/10/06/what-dorie-clarks-latest-book-can-teach-coaches-about-emotional-intelligence/#a7e7b4c6d976.

Tanaka, L. (2015). Language, gender, and culture. In F. Sharifian (Ed.), *The Routledge handbook of language and culture.* (pp. 100–112). New York: Routledge/Taylor & Francis Group.

Tannen, D. (1990). *You just don't understand: Women and men in conversation.* New York: Ballantine.

Thibodeau, P. H., & Boroditsky, L. (2011). Metaphors we think with: The role of metaphor in reasoning. *PLoS One, 6*(2), e16782. https://doi.org/10.1371/journal. pone.0016782. Retrieved from http://journals.plos.org/plosone/article?id=10.1371/journal.pone.0016782.

Tye-Williams, S., & Krone, K. J. (2015). Chaos, reports, and quests: Narrative agency and co-workers in stories of workplace bullying. *Management Communication Quarterly, 29*(1), 3–27.

US Department of Labor, Bureau of Labor Statistics. (2017, April 20) News Release. 'Employment Characteristics of Families — 2016.' Retrieved from: https://www.bls.gov/news.release/pdf/famee.pdf [Accessed 10/03/2017]

Williams, J. C., Phillips, K. W., & Hall, E. V. (2016). Tools for change: Boosting the retention of women in the STEM pipeline. *Journal of Research in Gender Studies, 6*(1), 11–75. Retrieved from http://repository.uchastings.edu/faculty_scholarship/1434.

Wolfe, J., & Powell, E. (2009). Biases in interpersonal communication: How engineering students perceive gender typical speech acts in teamwork. *Journal of Engineering Education, 98*(1), 5–16.

Wright, T. (2016). Women's experience of workplace interactions in male-dominated work: The intersections of gender, sexuality and occupational group. *Gender, Work and Organization, 23*(3), 348–362. Retrieved from http://library.pcw.gov.ph/sites/default/files/Wright-2015-Gender%2C_Work_%26_Organization.pdf.

Xie, Y., & Killewald, A. A. (2012). *Is American Science in Decline? Boston.* MA: Harvard University Press.

FURTHER READING

Acker, J. (2006). Inequality regimes gender, class, and race in organizations. *Gender and Society, 20*(4), 441–464. Retrieved from http://www.jstor.org/stable/27640904.

Aikens, K. A., Astin, J., Pelletier, K. R., Levanovich, K., Baase, C. M., Park, Y. Y., et al. (2014). Mindfulness goes to work: Impact of an on-line workplace intervention. *Journal of Occupational and Environmental Medicine, 56*(7), 721–731.

Armstrong, E. A., Hamilton, L. T., Armstrong, E. M., & Seeley, J. L. (2014). 'Good girls': Gender, social class, and slut discourse on campus. *Social Psychology Quarterly, 77*(2), 100–122.

Baldwin, R. G. (2009). The climate for undergraduate teaching and learning in STEM fields. *New Directions for Teaching and Learning, 117,* 9–17.

Bear, J. B., & Woolley, A. W. (2011). The role of gender in team collaboration and performance. *Interdisciplinary Science Reviews, 36*(2), 146–153.

Bednarek, M. (2012). Constructing 'nerdiness': Characterisation in the big bang theory. *Multilingua, 31,* 199–229.

Bhatt, M., Blakley, J., Mohanty, N., & Payne, R. (n.d.). How media shapes perceptions of science and technology for girls and women. FEMinc. Retrieved from https://learcenter.org/pdf/femSTEM.pdf.

Burgoon, J., Berger, C. R., & Waldron, V. R. (2000). Mindfulness and interpersonal communication. *Journal of Social Issues, 56*(1), 105–127.

Clancy, K. B., Lee, K. M. N., Rodgers, E. M., & Richey, C. (2017). Double jeopardy in astronomy and planetary science: Women of color face greater risks of gendered and racial harassment. *Journal of Geophysical Research: Planets, 122*(11), 1610–1623.

Croson, R., & Gneezy, U. (2009). Gender differences in preferences. *Journal of Economic Literature, 47*(2), 448–474.

Eddy, S. L., & Brownell, S. E. (2016). Beneath the numbers: A review of gender disparities in undergraduate education across science, technology, engineering, and math disciplines. *Physical Review Physics Education Research, 12.* Retrieved from https://journals.aps.org/prper/abstract/10.1103/PhysRevPhysEducRes.12.020106.

Ellemers, N. (2012). The group self. *Science, 336,* 848–852. Retrieved from http://sciencemag.org.

Elrod, S., & Kezar, A. (2015). Increasing student success in STEM. *Peer Review, 17*(2). Retrieved from http://www.aacu.org/peerreview/2015/spring/elrod-kezar.

Fernández-Berrocal, P., Cabello, R., Castillo, R., & Extremera, N. (2012). Gender differences in emotional Intelligence: The mediating effect of age. *Behavioral Psychology, 20*(1), 77–89.

Flaherty, C. (2014, Oct. 1). *Are STEM fields more gender-balanced than non-STEM fields in Ph.D. Production?* Inside Higher Ed. Retrieved from https://www.insidehighered.com/news/2014/10/01/are-stem-fields-more-gender-balanced-non-stem-fields-phd-production.

Goncalo, J. A., Chatman, J. A., Duguid, M. M., & Kennedy, J. A. (2015). Creativity from constraint? How the political correctness norm influences creativity in mixed-sex work groups. *Administrative Science Quarterly, 60*(1), 1–30.

Hall, C., Dickerson, J., Batts, D., Kauffmann, P., & Bosse, M. (2011). Are we missing opportunities to encourage interest in STEM fields? *Journal of Technology Education, 23*(1). Retrieved from http://scholar.lib.vt.edu/ejournals/JTE/v23n1/hall.

Handley, I. M., Brown, E. R., Moss-Racusin, C. A., & Smith, J. L. (2015). Quality of evidence revealing gender biases in science is in the eye of the beholder. *Proceedings of the National Academy of Sciences of the United States of America, 112*(43), 13201–13206.

Holmes, J. (2009). *'Discourse in the workplace literature review.' Language in the workplace occasional papers, Number 12.* Retrieved from http://www.vuw.ac.nz/lals/lwp.

Holmes, J., & Schnurr, S. (2006). Doing 'Femininity' at work: More than just relational practice. *Journal of Sociolinguistics, 10*(1), 31–51.

Hopewell, L., McNeely, C. L., Kuiler, E. W., & Hahm, J. (2009). University leaders and the public agenda: Talking about women and diversity in STEM fields. *The Review of Policy Research, 26*(5), 589–607.

Jaschik, S. (October 27, 2017). *In hiring for junior faculty positions, study finds bias against female candidates who have parters.* Inside Higher Ed. Retrieved from https://www.insidehighered.com/news/2017/10/27/hiring-junior-faculty-positions-study-finds-bias-against-female-candidates-who-have.

Kapidzic, S., & Herring, S. C. (2011). Gender, communication, and self-presentation in teen chatrooms revisited: Have patterns changed? *Journal of Computer-Mediated Communication, 17,* 39–59.

Kesckes, I. (2015). *'Language, culture, and context.' The Routledge handbook of language and culture* (pp. 113–128). New York: Routledge/Taylor & Francis Group.

Kinney, D. A. (1993). From nerds to normals: The recovery of identity among adolescents from middle school to high school. *Sociology of Education, 66*(1), 21–40.

Koch, S. C., Konigorski, S., & Sieverding, M. (2014). Sexist behavior undermines women's performance in a job application situation. *Sex Roles, 70*(3—4), 79—87. Retrieved from https://link.springer.com/article/10.1007/s11199-014-0342-3.

Lakoff, R. T. (2000). *The language war.* Berkeley: Univ. CA Press.

Lakoff, G., & Johnson, M. (1980). *Metaphors We Live BY.* Chicago, IL: University of Chicago Press.

Lee, J. S., & Anderson, K. T. (2009). Negotiating linguistic and cultural identities: Theorizing and constructing opportunities and risks in education. *Review of Research in Education, 33,* 181—211.

London, B., Rosenthal, L., Levy, S. R., & Lobel, M. (2011). The influences of perceived identity compatibility and social support on women in nontraditional fields during the college transition. *Basic and Applied Social Psychology, 33,* 304—321.

Lucha, E. (June 25, 2014). *Nipped in the bud: Women and STEM careers.* Linked.In. Retrieved from https://www.linkedin.com/pulse/20140625194514-79880216-nipped-in-the-bud-women-and-stem-careers.

Madsen, A., McKagan, S. B., & Sayre, E. C. (2013). Gender gap on concept inventories in physics: What is consistent, and what factors influence the gap? *Physics Education Research, 9*(2). Retrieved from https://arxiv.org/ftp/arxiv/papers/1307/1307.0912.pdf.

Mayo, C., & Henley, N. M. (Eds.). (1981). *Gender and nonverbal behavior..* New York: Springer-Verlag.

McDowel, J. (2015). Masculinity and non-traditional occupations: Men's talk in women's work. *Gender, Work and Organization, 22*(3), 273—291.

Mintz, S. (February 7, 2016). *Improving rates of success in STEM fields.* Inside Higher Ed. Blog. Retrieved from https://www.insidehighered.com/blogs/higher-ed-beta/improving-rates-success-stem-fields.

Morgan, S. L., Gelbgiser, D., & Weeden, K. A. (2013). Feeding the pipeline: Gender, occupational plans, and college major selection. *Social Science Research, 42,* 989—1005.

Moss-Racusin, J. F., Davido, V. L., Brescoll, M. J., Graham, & Handelsman, J. (2012). Science Faculty's subtle gender biases favor male students. *Proceedings of the National Academy of Sciences of the United States of America, 109*(41), 16474—16479.

Murray, P. A., & Syed, J. (2010). Gendered observations and experiences in executive women's work. *Human Resource Management Journal, 20*(3), 277—293.

Myers, K. K., Jahn, J. L. S., M.Gailliard, B., & Stoltzfus, K. (2011). Vocational anticipatory socialization (VAS): A communicative model of adolescents' interests in STEM. *Management Communication Quarterly, 25*(1), 87—120.

Nelson, D. J., & Brammer, C. N. (2010). *A national analysis of minorities in science and engineering faculties at research universities.* Retrieved from http://drdonnajnelson.oucreate.com/diversity/Faculty_Tables_FY07/07Report.pdf.

Nelson, R. G., Rutherford, J. N., Hinde, K., & Clancy, K. B. H. (2017). Signaling safety: Characterizing fieldwork experiences and their implications for career trajectories. *American Anthropologist, 119*(4), 710—722. Retrieved from http://onlinelibrary.wiley.com/doi/10.1111/aman.12929/full.

Oh, S. S., & Kim, J. (2013). Science and engineering majors in the federal service: Lessons for eliminating sexual and racial inequality. *Review of Public Personnel Administration, 35*(1), 24—46. Retrieved from http://journals.sagepub.com/doi/abs/10.1177/0734371x13504117.

Palmer, R. T., Davis, R. J., & Thompson, T. (2010). Theory meets practice: HBCU initiatives that promote academic success among African Americans in STEM. *Journal of College Student Development, 51*(4), 440—443.

Parson, L. (2016). Are STEM syllabi gendered? A feminist critical discourse analysis. *The Qualitative Report, 21*(1), 102—116. Retrieved from http://nsuworks.nova.edu/tqr/vol21/iss1/9.

Patten, E. (2014). *On equal pay day, key facts about the gender pay gap*. The Pew Foundation. Retrieved from http://www.pewresearch.org/fact-tank/2014/04/08.

Pisani, M. (2012). The impact of team composition and interpersonal communication on perceived team performance—A case study. *European Journal of Social Sciences, 35*(3), 411—430.

Price, J. (2010). *The effect of instructor race and gender on student persistence in STEM fields*. (Vol. 29, pp. 901—910) Economics of Education Review.

Ragusa, G. (2013). *Science literacy and textbook biases*. American Society for Engineering Education, Conference Proceedings. Retrieved from https://www.asee.org/file_server/papers/attachment/file/0003/3493/revised-Science_Literacy_and_Text_Book_Biases_ASEE_2013-FINAL.pdf.

Reynolds, J. A., Thaiss, C., Katkin, W., & Thompson, R. J., Jr. (2012). Writing-to-learn in undergraduate science education: A community-based, conceptually driven approach. *CBE—Life Sciences Education, 11*, 17—25.

Rivera, L. A. (2017). When two bodies are (not) a problem: Gender and relationship status discrimination in academic hiring. *American Sociological Review, 82*, 1111—1138.

Roberts, S. G., & Verhoef, T. (2016). Double-blind reviewing at EvoLang 11 reveals gender bias. *Journal of Language Evolution, 1*(2), 163—167. Retrieved from http://jole.oxfordjournals.org.

Rosenthal, H. E., Norman, L., Smith, S. P., & McGregor, A. (2012). Gender-based navigation stereotype improves men's search for a hidden goal. *Sex Roles, 67*(11—12), 682—699. Retrieved from https://link.springer.com/article/10.1007/s11199-012-0205-8.

Schnurr, S., & Chan, A. (2011). When laughter is not enough. Responding to teasing and self-denigrating humour at work. *Journal of Pragmatics, 43*, 20—35.

Sheltzera, J. M., & Smith, J. C. (2014). Elite male faculty in the life sciences employ fewer women. *Proceedings of the National Academy of Sciences of the United States of America, 111*(28)), 10107—10112. Retrieved from http://www.pnas.org/content/111/28/10107.full.pdf.

Tannen, D. (1986). *That's not what I meant! How conversational style makes or breaks relationships*. New York: Ballantine.

Tannen, D. (1996). *Gender & discourse*. New York: Oxford.

Tenopir, C., & King, D. W. (2004). *Communication patterns of engineers*. Hoboken, NJ: IEEE Press, Wiley-Interscience.

Thory, K. (2012). A gendered analysis of emotional intelligence in the workplace: Issues and concerns for human resource development. *Human Resource Development Review, 12*(2), 221—244.

Till, J., & Meares, H. (2015). *Women in science* (Vol. 31). Birmingham, AL: UAB Comprehensive Cancer Center (3).

Tsui, L. (2007). Effective strategies to increase diversity in STEM fields: A review of the research literature. *The Journal of Negro Education, 76*(4), 555—581. Retrieved from http://www.jstor.org/stable/40037228.

Ulriksen, L., Madsen, L. M., & Holemgaard, H. T. (2010). What do we know about explanations for drop out/opt out among young people from STM higher education programmes? *Studies in Science Education, 46*(2), 209—244. Retrieved from https://doi.org/10.1080/03057267.2010.504549.

US Department of Commerce, Economics and Statistics Administration. (August 2011). *Women in STEM: A gender gap to innovation*. Retrieved from http://www.esa.doc.gov/reports/women-stem-gender-gap-innovation.

Valian, V. (2007). Women at the top in science—and elsewhere. In S. J. Ceci, & W. M. Williams (Eds.), *Why aren't more women in science: Top researchers debate the evidence* (pp. 27—37). Washington, DC: American Psychological Association.

Velegol, S. B. (May 5, 2016). *The importance of women in academe asking bold questions*. Inside Higher Ed. Retrieved from https://www.insidehighered.com/advice/2016/05/05/importance-women-academe-asking-bold-questions.

Wagner, C. (2016). Rosalind's ghost: Biology, collaboration, and the female. *PLoS Biology*, *14*(11). Retrieved from http://journals.plos.org/plosbiology/article?id=10.1371%2Fjournal.pbio.2001003.

Watkins, J., & Mazur, E. (2013). Retaining students in science, technology, engineering, and mathematics (STEM) majors. *Journal of College Science Teaching*, *42*(5), 36–41.

Waytz, A., Dungan, J., & Young, L. (2013). The Whistleblower's dilemma and the fairness-loyalty tradeoff. *Journal of Experimental Social Psychology*, *49*, 1027–1033.

Williams, W. M., & Ceci, S. J. (2015). National hiring experiments reveal 2:1 faculty preference for women on STEM tenure track. *Proceedings of the National Academy of Sciences of the United States of America*, *112*(17). Retrieved from www.pnas.org/cgi/doi/10.1073/pnas.1418878112.

Wolfe, J. (2012). Communication styles in engineering and other male-dominated fields. In B. Bogue, & E. Cady (Eds.), *Applying research to practice resources*. Retrieved from http://www.engr.psu.edu/awe/ARPresources.aspx.

Wolfe, J., & Powell, E. (2006). Gender and expressions of dissatisfaction: A study of complaining in mixed-gendered student work groups. *Women and Language*, *29*(2), 13–20.

Zell, E., Krizan, Z., & Teeter, S. R. (2015). Evaluating gender similarities and differences using metasynthesis. *American Psychologist*, *70*(1), 10–20.

Zeng, X. H. T., et al. (2016). Differences in collaboration patterns across disciplines, career stage, and gender. *PLoS Biology*, *14*(11). Retrieved from http://journals.plos.org/plosbiology/article?id=10.1371/journal.pbio.1002573.

INDEX

'*Note:* Page numbers followed by "f" indicate figures, "t" indicate tables.'

A

Abstract worker, 13
Airbags, 83–84
American Civil Rights Movement, 25
Argument, 16
Assertiveness, 84
Automobiles mechanics, 15

B

Baby Boom (film), 71–72
Behavior, 63
Big Bang Theory, 23
Brainstorm, 21
Breastfeeding, 32
Business attire, 76

C

Civil Rights Movement, 25
Communication, 79–80. *See also*
 Workplace(s)—communication
 communicating as professionals in
  STEM
  Jane's story, 82
  Lori's story, 83–86
  Mary's story, 82
  Sadie's story, 80–82
 culture and, 15
 gender, 59–61
 mindfulness in, 80
 nonverbal, 17
 styles, 61, 66
Creative jokesters, 47
Crime, 16–17
"Culturally neutral" ways, 73
Culture, 16
 and communication, 15

D

Dress for Success (Mollow), 71–72
Dressing for success
 business attire, 76
 personal branding, 76–77
 physical bodies, 73
 gender and dress, 74–75
 women and professional attire, 71
 women in STEM, 77

E

EQ. *See* Emotional intelligence (EQ)
Emotional intelligence (EQ), 44–45
Empathic listening, 80

F

Fairness, 81–82
Family Leave Act, 85–86
Feelings, 30, 52
Feminine communication styles,
 64–65
Feminine interactional styles, 61f
Fletcher's framework, 47–48

G

Gender(ing), 13–14, 64–65
 communication, 59–61
 nurse, 6–13
 organizational, 13
 within organizational structure, 13
 roles, 17–18

I

Internalization of societal beliefs,
 27–28
Interpersonal communication, 79

J

Job sharing, 85–86

L

Language, 17
Leaders, 53
    leader-like qualities, 36–39
    organizational, 86
    in STEM fields, 66
Leaky pipeline, 1–2
    bachelor's, master's, and doctor's
        degrees conferred, 3t–5t
    brainstorm, 21
    Elizabeth's story, 14–21
    national science foundation enrollment
        status of S&E graduate students,
        8t–12t
    STEM degrees conferred by US
        postsecondary institutions, 7t
LGBTQ community, 17–18
Living incarnation of stereotypes,
    36–39
Locker room humor tests, 47
Long-running stereotype, 27–28
Loudest voice phenomenon, 63

M

Masculine communication styles,
    63–65
Masculine interactional styles,
    61f
Mentors, 35–36, 55–56, 70, 77,
    83
Metaphors We Live By, 16
Mindfulness, 53
    in communication, 80
Miscommunication, 62
Mutual empowering, 46

N

Name harassment and disrespect, 84
National Center for Education Statistics
    (NCES), 2
National Science Foundation, 2–6
Nonverbal communication, 17
Nurse gender, 6–13

O

Off-record mentoring, 47–48
Office small talk inclusion
    Amy's story, 43–45, 49, 52–56
    Cara's story, 49
    informants comments on small talk,
        50–52
    Jeanna's story, 45
    Jessica's story, 45–49
    Susan's story, 42–43
Organizational/organizations
    gendering, 13
    leaders, 86
    structure jobs, 6–13

P

Peers, 35–36
Personal branding, 75–77
Plain language, 16–17
Power dressing model, 72
Pregnancy, 35, 73
Preserving relational practice, 46
Professional women, 75

Q

Qualifiers, 62

R

Regardless of gender, 74–75
Relational practice, 46, 48
Remember the Titans (movie), 46–47

S

Self-achievement asserts, 47
Self-achieving, 46
Sexism, 81
Silent Spring (Carson), 1–2
Small talk, 41–43
    informants comments, 50–52
Social talk, 42–43
Solitary journey, 55
Speech, 15, 18, 61–62
STEM
    fields, 6, 14, 79
        women in, 1–2
    professionals, 56

professions, 20—21
stereotypes, 23
    Anne's story, 28
    Greta's story, 30—31
    Jessica's story, 28—32, 34—39
    Juanita's story, 34
    Karen's story, 32—33
    percent and average composite score, 26t
    Tanya's story, 33—34
    women in, 1—2, 77, 85
    workplaces, 48—49, 85—86
Stereotypes, 24—25
    living incarnation of, 36—39
    threat, 25—27
Storytelling, 15, 43
Stubbornness, 29—30

T

Team building exercises, 86
Teams creation, 46
Traditionally male workplaces, 48

U

Unencumbered white male, 13
US Department of Labor, 36, 37t—38t

W

Watzlawick's theory, 79
Whoever Tells the Best Story Wins (Simmons), 43
Women, 65—66
    of color, 54, 72—73
    dressing, 71
    in STEM, 1—2, 77, 85
    in workplace, 75
Women's Dress for Success Book, The (Mollow), 71—72
Workplace(s), 62
    bullying, 54
    communication
        Ann's gems, 69—70
        Ann's story, 68—69
        continuum of masculine and feminine interactional styles, 61f
        Deborah's story, 59—67
        gender in the workplace, 73
        Jade's story, 68
        as village of men, 72, 76

Printed in the United States
By Bookmasters